Michael Jam[illegible]

DOMESTIC AIRLINE REGULATION

The Australian Debate

CIS Research Studies in Government Regulation 1

DOMESTIC AIRLINE REGULATION

The Australian Debate

Michael G. Kirby

Australian National University

CIS

THE CENTRE FOR INDEPENDENT STUDIES

1981

First published February 1981 by

The Centre for Independent Studies

National Library of Australia

Cataloguing in Publication Data

Kirby, Michael G. (Michael Gerard), 1953-
Domestic airline regulation

Bibliography
ISBN 0 9596485 9 3.

1. Airlines, Local Service - Australia. 2. Aeronautics, Commercial - Law and legislation - Australia.
I. Centre for Independent Studies (Australia). II. Title.
(Series : CIS research studies in government regulation ; 1).

387.7'4042

PRINTED AND PRODUCED BY:
LINDSAY YATES & PARTNERS PTY. LTD.
& PROMAIL PRINTING GROUP PTY. LTD.

Contents

The Author

Michael Kirby completed his B.Ec. at the University of Sydney in 1974. Having been a Treasury Cadet (Economics) in his final honours year of study, he moved to Canberra and worked in the Treasury during 1975 and 1976, after which he joined the Bureau of Agricultural Economics (BAE). He was awarded a Commonwealth Public Service Postgraduate Scholarship to undertake a M.Ec. at the Australian National University during 1978. Following a brief return to the BAE he took leave-without-pay from the Commonwealth Public Service and in 1979 commenced research towards a Ph.D. on the economic regulation of Australia's domestic air transport. Michael Kirby has had articles on the airline industry and other topics published or forthcoming in the *Australian Journal of Management*, *Transportation Research*, *Economic Record* and *Australian Economic Papers*.

Acknowledgements

I would like to thank the many readers of earlier drafts of this monograph for their helpful comments, suggestions and encouragement. In particular, I must acknowledge the substantial efforts of Peter Swan, Christopher Findlay, Christopher Hackett, Mark Grenning, Lynette Kirby, Robert Albon and the anonymous referees. I am also grateful to the various industry and other sources from which I was able to obtain some of the information and raw material necessary for this study, and to the Centre for Independent Studies for giving me the opportunity to publish the findings of my research. Of course, I alone accept full responsibility for the final contents and viewpoints contained in this publication.

Preface

Government regulation of an increasing level of economic activity has become a feature of modern life. The traditional justification for economic regulation is that intervention is needed to correct so-called 'market failure' and is thus intended to serve the public interest. There is, though, some degree of argument as to the extent of market failure. This argument has become more current given the evidence of the failure of government to deal in any satisfactory way with some of the problems it has set out to cure through intervention. Perhaps markets do fail, but it may be that the costs of 'government failure' are far greater.

Some argue that the failure of the regulatory process to achieve its ends in correcting the performance of the market is sound enough reason on grounds of efficiency for abandoning regulation altogether. Others argue, with increasing force that **any** interference with the voluntary market process inevitably leads to unwanted side-effects.

There are aspects to government regulation which can never be fully appreciated - often because the effects are unseen. These include the burden of costs of compliance with the regulations upon individuals and companies (and their customers).

More fundamentally, regulation leads to a serious disruption of the market process itself, resulting in a myriad of foregone innovations and lost opportunities due to the stifling nature of many regulations. This disruption of the market process has another aspect. An unregulated market generates a discovery process under competition which regulation hinders. The competitive market process is a way of discovering who can do a job best in given circumstances. The person (or company) who does best is not the same in all circumstances, but will change as conditions change. Regulation acts as if to decide the winner in the competition stakes in advance and to prevent those who may wish to join the race from competing on anything like equal terms. The costs to efficiency and consumer welfare may then be very high.

In recent times, the pendulum that was swinging towards greater interference in the economy has slowed.

Serious attempts have been made to determine if there are other reasons why regulation exists. If regulation is instituted in the name of the 'public interest' and it can be shown that the public interest is not being served, then perhaps other explanations are called for. If in fact regulation does operate entirely to correct market failure (bearing in mind that there are strong views about whether regulation could ever do this satisfactorily), then there may be little argument about the need for regulation. However, on observing just some incidences of economic regulation, a number of points which go counter to the market failure/ public interest rationale, seem to arise.

For instance, regulation frequently acts not to improve the efficiency of markets, but to affect a redistribution of income towards those groups subject to regulation, and away from, and at the expense of, the larger body of consumers. As well, regulators can fall captive to the groups which they are charged with regulating and so end up serving a small and privileged group of the regulated at the expense of new entrants to the industry.

There are further undesirable features of regulation - too many to be dealt with in a short preface. The Centre for Independent Studies has long been concerned with the effects of government economic regulation, and indeed of regulation of all individual activity. In establishing this new series of *Research Studies in Government Regulation*, the Centre will employ some of Australia's ablest social scientists to undertake specific studies. By examining regulation in the context of particular cases, general principles with which to deal with broader issues of regulation will be established.

This first study by Michael G. Kirby, an economist from the Australian National University, is a substantial piece of research into the regulation of an industry which exhibits some of the more notorious side-effects of regulation.

To Australians, the existence of the Two-Airline Policy, and regulation of the airways in general, have come to signify the problems of regulation of industry in a highly visible way. The presence of a government-owned airline has not helped matters. The continual public squabbling over some aspect or other of government policy as it affects the domestic airline industry has not only diverted the government away from its more legitimate functions, but has made sure that the machinations of airline regulation are continually dragged into the daily press. Is the fact that an airline serves sandwiches or chicken really worthy of front-page headlines? In short, the whole issue has become one of unnecessary prominence, reflecting the inordinate social cost of making such decisions in a political process.

The debate over the regulation of Australia's airlines has been loaded with contradiction, political expediency and unsupportable economic argument. Mr Kirby has helped to clarify the issue with his thorough sifting of a broad cross-section of opinion. He carefully analyses the many propositions put forward in the defence of regulation and skilfully shows how so often its champions fail to survive even the most basic economic test. On surveying this study it is hoped that the reader may gain some insights into the complex issue of economic regulation in the general sense, as well as a better understanding of the development of what has become in a short space of time a policy which seems to bestow no credit on those whose aim is to serve the 'public interest'.

In the end, the reasons for instituting regulation may be at odds with the 'public interest' rhetoric with which it is clothed. Rigorous empirical analysis allied with the insights of theory may bring more to bear on the issue than is now obvious. Out of it all, and this is the purpose for the establishment of this series of studies on regulation, the viability of a competitive market may be maintained and the opportunities of service to the consumer enhanced.

Adam Smith long ago saw the virtues of competition. He delivered many warnings, but perhaps none was more perceptive than:

> People of the same trade seldom meet together, even for merriment and diversion, but the conversation ends in a conspiracy against the public; or in some contrivance to raise prices. It is impossible indeed to prevent such meetings, by any law which either could be executed, or would be consistent with liberty or justice. But though the law cannot hinder people of the same trade from sometimes assembling together, it ought to do nothing to facilitate such assemblies, much less to render them necessary.

The Centre for Independent Studies is pleased to publish this book, which it feels makes a significant contribution to the debate on airline policy and regulation generally. However, the conclusions of its author remain his alone and cannot be considered to be those of the Centre's Directors, Trustees, Advisers or officers.

Greg Lindsay

Introduction

> . . . a world of half-measures, complicated compromises, and political 'realities' created by fervent repetition of untruths . . . [Levine (1975, p. 703)] *

The debate on the regulation and possible deregulation of the Australian civil aviation industry consists of many complex, interrelated and sometimes conflicting arguments, which are often presented in the context of subjective value judgements. This study critically and comprehensively reviews the many arguments presented throughout the development of Australian aviation policies, and is intended to serve as a guide to interested participants in, and observers of, this airline regulation debate. While the above quotation refers explicitly to the U.S. experience, it will soon become apparent that it is also applicable to Australia.

The study is directed mainly towards the regulation of the Australia's domestic air transport industry.[1] There is also a tendency to concentrate on the provision of trunk route services, although the analysis is not limited in its application to these particular services. This concentration reflects the overwhelming importance of the Two-Airline Policy for domestic civil aviation. However, while there is a definite domestic setting, much of the discussion also has direct

---

* Due to the research emphasis of this study and its extensive use of quotations and other supporting references, a combination of the Harvard system (author, date) and standard footnoting has been used. For those unfamiliar with the Harvard system, an example such as: Levine (1975, p. 703) refers to a 1975 study of Levine's as listed in the references at the end of the book.

1 The principal features of this regulatory system are detailed in Appendix 1.

relevance to the regulation of Australia's international aviation; though, of course, the latter has many unique features which are not considered here.

This book is intended to be an examination of a genuine and continuing debate, not merely a list of straw-man arguments. To achieve this purpose and to present with clarity the many different points of view, the study makes extensive reference to the many previous contributors to discussions of civil aviation policy. In addition, the empirical experience of the airline industry, mainly in Australia and the U.S., is used wherever possible as evidence when assessing these various points of view.

The study also seeks to analyse the Australian airline regulation debate within a sound economic framework. The discipline of economics offers criteria for the choice of industry policy likely to increase the community's welfare. In particular, it provides two necessary conditions for an economic efficiency justification of government intervention with market forces. Firstly, one must demonstrate the existence of market failure where the market outcome differs from the socially optimal one, and hence where there exists the potential to increase efficiency through government regulatory actions. Secondly, the benefits from any proposed market correction policy must exceed its costs. Hence a thorough analysis of the airline regulation issue must involve a consideration of the likely extent of both market failure and regulatory failure. A constantly occurring theme throughout such an analysis is the effect on the economic performance of an industry of the incentives facing its participants, i.e. firms, consumers and regulators.

Chapters 1 to 5 scrutinise the various arguments regarding alleged market failure in the provision of airline services. Chapters 1 and 2 examine the industry structure and patterns of behaviour which are likely if open market competition among airlines were permitted. These chapters consider the likelihood that airline competition would be destructive and/or wasteful, and the alternative possibility that the industry would tend towards monopoly and a lack of competitive behaviour. One of the special features of the Australian air transport industry is the existence of a government-owned firm. Its rationale and impact are discussed in Chapter 3. The important question of air safety and its relevance to the economic regulation of airlines are considered in Chapter 4, while Chapter 5 examines some social objectives which are frequently mentioned in the civil aviation context.

The following two chapters examine some issues relating to government failure in Australian air transport. The debate on the Two-Airline Policy is surveyed in Chapter 6. This chapter discusses the need for policy reform and broadly considers some of the available regulatory options. Chapter 7 reminds the reader of some of the inherent tendencies of the regulatory and competitive processes and suggests that a more useful explanation of current aviation policies may be found in the private interest theory of regulation rather than in the public interest theory which underlies a concern with market failure.

The study's Conclusion briefly summarises the implications of this analysis of the airline regulation debate for policy reform and speculates on its likelihood.

1
Competition Among Airlines: Too Much?

I. DESTRUCTIVE COMPETITION

Chaos and instability

The claim has often been made that permitting open market competition would lead the airline industry into a permanent state of chaos and instability. Thus:

> Competition can exist to such a degree as to destroy all the several airlines, . . . [2]

The poor financial returns earned by some airlines during the post-war decade were frequently attributed to the effects of competition. Certainly Government policy of the time was dominated by a fear that a competitive airline industry would be inherently unstable. Senator Paltridge (Minister for Civil Aviation, 1956 - 1964) remarks:

> It is not very rewarding in these circumstances to find a financial crisis developing in the industry every few years . . . it is essential that we get a stable pattern for future development which holds some prospect of the industry finally emerging as a self-sufficient arm of our transport services.[3]

The Government's response to achieve stability consisted of tighter restrictions on competitive forces in the industry. Import controls were used to prevent the entry of other operators, rationalisation procedures for airline services were strengthened, and measures to control capacity were introduced. A further example of the anti-competitive nature of the Government's policy is provided by Brogden

---

2 Bland, Commonwealth of Australia, *Parliamentary Debates* (hereafter *Hansard*), *House of Representatives* (hereafter *HR*), 29 October 1952, p. 3885.

3 *Hansard*, *Senate* (hereafter *S*), 3 September 1957, p. 82.

(1968, p. 151) who reports that under the Cross Charter Agreement 1960 the airlines agreed not to introduce or propose any reduced fare services. The Two-Airline Policy (hereafter TAP) allegedly grew out of post-war chaos,[4] and, as a result of this policy,

> . . . the industry has been spared for many years the periodic economic crises that bedevilled it for so much of its history. [Ansett (1976, p. 54)]

The argument that competition must be restricted in order to prevent a collapse of the airline industry is still an important influence on regulatory policy. Robinson asserts:

> . . . whilst there is a competitive situation in certain respects obviously there must be basic things that have some common ground or the whole industry would be in chaos . . .[5]

Similarly, Hocking (1979a, p. 72) warns that if the possibility of greater price differentiation is allowed in the industry:

> Some degree of co-operation between airlines will probably be essential, as widely differing prices between airlines, even if only temporary, could be destabilising.

While most authors agree that the TAP has maintained stability in the airline industry [e.g. Forsyth (1979, p. 66)], it is not clear that the industry would have been subject to destructive competition and instability in the absence of such a policy. It is also not clear that the competition-and-chaos scenario is the most accurate interpretation of events in the post-war period.

Destructive pricing and airlines

Kahn (1971, p. 173) identifies two prerequisites for des-

---

4 Ansett (1972, p. 39). Sir Reginald Ansett was founder, Chairman and Managing Director of Ansett Transport Industries (ATI).

5 Robinson, Commonwealth of Australia, *House of Representatives, Select Committee on Tourism* (hereafter *Hansard, SCT)*, 8 February 1977, p. 799. Robinson was subsequently Minister for Finance.

tructive pricing to occur in an industry. Firstly, fixed costs must be a large proportion of total costs; and secondly, the industry must be characterised by sustained and recurrent periods of excess capacity. When these two conditions are satisfied the firm is operating on the downward-sloping portion of its short run average cost curve and its scale of operations is too large for its level of demand. Since marginal costs are less than average costs in such circumstances, competition tends to depress prices towards the former and firms incur losses. When this argument is applied to the aviation industry we have the 'sacred truth' referred to by Levine (1965, p. 1423): without restrictions on competition, airlines would indulge in a frenzy of below-cost selling which would ultimately ruin them all.

Does the airline industry possess these prerequisites for destructive pricing? Supporters of regulation argue that the industry is characterised by a substantial degree of fixed costs.[6] However, there is a strong consensus among economists that the ratio of fixed to variable costs is low for airline operations.[7] Attention should also be directed towards the second necessary condition for destructive pricing which requires that the firm is unable to readily adjust its scale of operations to its most efficient level. However, it appears that airlines are capable of adjusting their scale fairly rapidly:

> . . . the airline industry is very flexible and will gear up swiftly to meet public demand when growth returns to the industry. [Ansett (1976, p. 53)]

Many fixed costs can be quickly altered since an airline's capital assets seem quite mobile.[8] For example, aircraft can be bought and sold or leased without great difficulty and readily shifted from one market to another. Computer systems, and perhaps even maintenance and engineering faci-

---

6 E.g. Goodrich (1960, p. 229), Chippindall (1965) and Brenner (1975, p. 798). Sir Giles Chippindall was Chairman of Trans-Australia Airlines (TAA), 1959-1966.

7 E.g. Smith (1978, p. 58), Grenning and Coat (1979, p. 12) and Keplinger (1976, p. 195). The validity of this conclusion is likely to have been enhanced by the rapidly rising fuel prices of recent years.

8 The nature and mobility of the fixed cost assets of airline operators are also important considerations for barriers to entry into the industry [see Chapter 2].

lities, can be transferred with relative ease to other firms for use in the same, or often a different, industry. These considerations of the nature of airline costs indicate that destructive pricing and chaos would be unlikely to result from open market competition in the air transport industry.

This assessment is supported by the U.S. experience with competitive airline markets. Lloyd-Jones (1975, p. 815) claims that when the Civil Aeronautics Act was introduced in 1938 the industry was in chaos with most carriers experiencing serious financial problems. However, according to Keplinger (1976, p. 194), this alleged chaos was 'more a fiction . . . than a reality.'[9] The below-cost bidding for mail contracts which occurred at the time reflected a desire to maintain a route monopoly on passenger traffic, as well as the knowledge that any losses would be covered by government subsidy and that further protective legislation was likely to be forthcoming. Thus any cutthroat competition which then existed could be more accurately attributed to government regulation under the Mail Act, rather than an inherent tendency towards destructive pricing. An examination of the Californian intrastate airline industry, which was virtually free of economic regulation prior to 1966, also reveals that airline competition need not be ruinous or chaotic [see Levine (1965, p. 1430)]. Forsyth and Hocking (1978, p. 22) summarise this experience by noting that free entry did not lead to instability, either in terms of the routes served by the different carriers or in terms of increased variability of profits.

It is important not to confuse the phenomenon of destructive competition with the normal competitive process. When excess capacity occurs (through, say, a decline in demand or the entry of a new firm into the industry) and a firm is operating on a scale greater than its most efficient level, short run losses and reduced investment in or exit from the industry represent the usual competitive adjustment process towards a new long run equilibrium position. This process encourages firms to operate at the appropriate scale and enables the most efficient firms to continue production. However, existing firms are likely to define instability in terms of the possibility of themselves suffering a declining market share or going out of business. Similarly, as Kahn (1971, p. 173) notes, all competition is potentially destructive to the equity of firms subjected to it since they may be replaced by more efficient operators. Consequently, it is

---

9 Levine (1965) and Miller (1975) also support this view.

typical for these firms to refer to any price competition in their industry as destructive, excessive or cutthroat.

Thus, while open competition among airlines may increase instability in the sense that the relative market fortunes of particular operators may vary over time, it does not necessarily follow that services to the consumer are unstable or disrupted. But is it possible for these competitive forces to have an adverse impact on consumer welfare?

Declining quality standards

One argument suggests that competitive forces will lead to a lower quality service since prices, schedules and other product characteristics would be widely fluctuating. Furthermore, in such circumstances consumers would be unable to evaluate this range of service qualities since they lack the necessary technical expertise. However, this 'bewildered consumer' outcome is unlikely. Firms have an incentive to minimise needless changes in prices and schedules since this strategy lessens transactions costs to the consumer and hence increases demand [see Miller (1975, p. 696)]. The competitive U.S. intrastate markets have provided no evidence of deteriorating quality standards through wildly fluctuating product characteristics in an open market setting. While some proliferation of different fares and services available might be expected immediately following the deregulation of airlines, this process is likely to be relatively short-lived; this has been the recent experience in the U.S. [see Kahn (1979, p. 11)]. A wider range of price/quality options which may be offered in the longer term in a competitive airline market would reflect closer satisfaction of consumer demands, and expert advisers (e.g. travel agents) could ease any difficulties posed by increased complications facing consumers. Finally, the argument at best indicates the possible need for policies designed to directly attack the supposed quality problem (e.g. certification of standards or fair advertising laws) rather than comprehensive market entry restrictions.

Collapse of airline investment

Competition is also alleged to have a harmful effect on the air transport industry through its impact on investment. Brenner (1975, p. 812) considers that it would be impossible to finance the large capital investments of airlines if free entry were permitted. Lloyd-Jones (1975, p. 825) claims that this impact will also hinder technological progress in the industry. The resultant need for route security and stability as a prerequisite to finance airline investments has also been a popular argument in Australia. Senator Cotton (Minister for Civil Aviation, 1969-1972) argues this case,[10] while Ansett (1965, p. 13) notes:

> Stability is essential in an industry spending millions on future equipment needs . . .

TAA holds a similar view:

> The stable economic environment has provided a sound base for the large commitments necessary for fleet purchases.[11]

However, the argument is misleading. While every firm would undoubtedly enjoy guaranteed profits, simple observation indicates that risk capital is forthcoming in many fields, including aviation, in the absence of such guarantees or security. Thus one would not expect investment in the airline industry to collapse in the presence of open competition. Furthermore, the claim that competition would result in destructive pricing and falling investment appears to be internally inconsistent. The losses from excess capacity which allegedly lead to a fall in investment have resulted from too much previous investment. This process cannot be sustained indefinitely. At best, the argument is a description of the period of adjustment to the optimal scale of operations; at worst, the argument is illogical. Finally, it is more probable that competitition would encourage technological progress, since those firms using the current best-technology techniques would be the ones most likely to prosper.

Irrational behaviour

Many of the claims made in support of airline regulation and examined in this study are likely to be valid only if airline firms behaved in an irrational manner. While it may be sensible and efficient to operate existing capital assets at short run marginal costs when faced with competitive

---

10 *Hansard*, *S*, 29 August 1972, p. 476.

11 *Hansard*, *SCT*, 8 February 1977, p. 706.

pressures, it would be irrational to invest in such an industry knowing that prices would not fully cover total costs of operation. Although Rasenberger (1975, p. 857) claims:

> . . . there is an allure to running an airline that seems to drive otherwise sensible people toward economic self-destruction . . . ,

economists would tend to reject this notion of irrationality. It also appears that Australian airline operators have not been as quick to claim self-irrationality!

Several comments are relevant. Firstly, the economic theory of competitive markets does not predict that failed investments will not occur. Competition involves rewarding good managerial decisions and penalising bad. Thus one should not confuse mistakes with irrationality. It follows that since one must allow for mistakes, which need not imply irrationality, any analysis of the performance of the industry should be primarily concerned with average behaviour over time and not with the financial fortunes of particular airline companies.

However, more mistaken investments might be expected in the period immediately following the removal of restrictive industry regulation. This could be due to a relative lack of knowledge regarding demand and supply conditions fostered by conservative regulation limiting the industry's operations to particular segments of the market. Even existing operators may not be certain of these conditions, except for those segments of the market for which they currently cater. It is likely that this situation would be quickly stabilised by the natural learning process and related market adjustments. This point is illustrated by the experience of the U.S. airline industry following its recent deregulation. The U.S. policy liberalised market entry, gave increased freedom to reorganise airline networks and offered prospects for a rise in competitive behaviour.[12] Several of the major operators, especially Braniff, reacted to this less constrained environment with overly ambitious expansion efforts. Subsequent losses have led to various merger negotiations, e.g. between Braniff and Eastern Air Lines.[13]

Finally, the often noted low profitability and marginal nature of the airline industry may merely reflect the level of normal profits required by owners of airline assets.

---

12 See Institut du Transport Aerien [ITA] (1979).

13 See *Time*, 4 August 1980, p. 49.

Alternatively, low profits and perhaps a relatively high variability of returns may to a certain extent have been compensated for by a truncation of the range of possible rates of return, i.e. the virtual guarantee of protection from bankruptcy.

II. WASTEFUL COMPETITION

The problem of wasteful competition

Proponents of airline regulation pursue a slightly different direction when the concept of wasteful competition is introduced. It is often feared that a freely competitive airline market will be characterised by wasteful competition through increased service quality resulting in higher prices. Regulatory policies are thus supposed necessary to restrict this competition and to protect the consumer. However, it seems clear that fallacious reasoning underlies much of this fear.

One line of argument suggests that since we observe cost-increasing competition within the regulated environment,[14] the situation would be even worse if the market were fully competitive. Hence more, not less, regulation is needed. However, this conclusion does not necessarily follow. It is the inability to compete with respect to prices that encourages firms to engage in non-price competition. Hence a viable alternative approach to reduce any observed wasteful competition is to remove price and entry restrictions. Kahn (1971, p. 209) summarises the economic principle involved:

> . . . when limitations are placed on price competition, but market conditions are such as to make continued interfirm rivalry likely, the consequence will be accentuation of service competition . . . Specifically, they will be inclined to improve service in one way or another, until their marginal costs, inflated by the service improvements, are equated to price.

Thus wasteful quality competition can often be traced directly to price regulation. In fact, since there is a direct

---

[14] For example, the 'free' drinks and meals which Ansett Airlines of Australia (AAA) initiated in the early months of 1980 are estimated to cost the operator at least $5m per annum [*The Canberra Times*, 7 May 1980].

positive relationship between price and quality when rivalry among firms reduces profits to normal levels, the observation of 'excessive' non-price competition indicates that the regulated price is set 'too high'. There are three policy choices: further restrictions in an attempt to remove all possibility of competitive behaviour and to enable firms to enjoy supernormal profits at the current regulated price; set a lower regulated price, implying less quality competition; or remove price and entry restrictions and permit market forces to determine the price/quality outcome.

The choice among these approaches depends upon an assessment of whether the regulatory authority or the market mechanism is better able to achieve the price/quality option preferred by consumers. Many economists would favour the objective assessment of price and quality offered by a fair market test. If the market indicates that consumers prefer a higher price/higher quality combination than is presently provided, this can hardly be interpreted as wasteful competition. The likelihood of regulators being better able to provide maximum consumer satisfaction diminishes rapidly with the knowledge that the observed inefficiences are due to current regulatory policies and with the recognition that the demand for air services is not homogeneous, yielding the possibility that many price/quality options may be in demand.

Some advocates for regulation claim that price competition is not feasible in airline markets. Brenner (1975, p. 803) erroneously suggests that competitive pricing behaviour implies different prices among airlines. In fact, the process leading to uniform prices, which he describes, is price competition. He also argues that a typical operator, when faced with a competitor offering a lower price/lower quality product, would match the price cut but maintain the higher quality and thus capture the entire market. However, in a competitive market this strategy involves short run losses. Such predatory pricing can be successful only if the firm is able to earn supernormal profits in later periods to offset the current losses. In turn this possibility depends on the existence of entry barriers. The question of entry barriers is discussed in greater depth in the following chapter; it is sufficient here to suggest that predatory pricing is unlikely to be prominent in competitive airline markets.

The remainder of this section examines three areas of concern which are closely associated with the notion of wasteful competition in the air transport industry: the provision of excess capacity, the airline equipment race and the occurrence of parallel schedules. Each of these cases illustrates the general principles discussed above.

Excess capacity

A persistent fear expressed in the airline regulation debate is that open competition will force the industry into a state of chronic overcapacity. It is alleged that the normal market forces of supply and demand do not apply in the airline industry, resulting in the absence of any self-adjusting equilibrium and a built-in tendency towards overcapacity [e.g. Brenner (1975)]. In particular:

> the product of an airline can easily become excessive in supply as an operator's competitive position is positively related so directly to the quantity of product.[15]

Hence, it is argued:

> . . . careful capacity regulation is a prerequisite for the survival of economically viable air transport operations.[16]

The capacity determination procedures of Australia's TAP ensure control of the aircraft capacity available to be offered by the two major operators. Chippindall (1965) foresees the 'creation of waste in unused capacity' if this policy were to be abandoned. Similarly, when recommending retention of the principle of capacity control, DOT (1979a, p. 59) concludes:

> . . . abolition or major relaxation of capacity controls could produce undesirable effects similar to those experienced overseas.

The overseas experience referred to is probably that of the U.S. where, under Civil Aeronautics Board (CAB) regulation, the achieved load factor was consistently less than the target load factor. Consumers are alleged to suffer from the

---

15 Department of Transport (DOT), *Annual Report*, 1975/76, p. 29. Civil aviation became the responsibility of DOT in November 1973 when the Department of Civil Aviation (DCA) was amalgamated with the previous DOT. Since its formation in 1938 DCA had been responsible for civil aviation.

16 DOT, *Annual Report*, 1974/75, p. 19.

competitive overcapacity situation through the resultant upward pressures on costs and hence prices. Thus:

> It was . . . the consideration of the public interest that we should not allow a free-for-all situation which in our view would force prices up rather than down.[17]

A major problem with the excess capacity scenario presented above is that it pays insufficient attention to the ceteris paribus assumptions implicit in the argument. While it seems clear that, if all other factors remain unchanged, a more frequent schedule would offer an operator an advantage over his rivals, in a competitive market this strategy must be accompanied in the long run by a higher price. In this more realistic setting a higher price/higher quality product will not necessarily be preferred by consumers. The tendency towards overuse of frequency competition can be attributed to a lack of price competition, i.e. to the control of fares. In the U.S. case the regulated prices, which were set by the CAB on the basis of a specified rate of return at a target load factor, implied supernormal profits and, not unexpectedly, led to non-price competition, mainly in the form of increased scheduling.[18] If the capacity offered is judged to be 'excessive', the blame lies with the authorities who set the price 'too high'. Hence, while Brenner (1975, p. 802) claims:

> . . . simple logic would suggest that, if the existing level of competition has created too many seats, totally unrestrained competition would create still more empty seats . . . ,

the above discussion indicates that this logic, while simple, is also incorrect.

The observed experience of competitive airline markets also indicates little need to fear chronic overcapacity. ITA (1979) reports on the impact of the recent U.S. deregulation

---

[17] Freeland, *Hansard*, *SCT*, 17 August 1978, p. 5591. Freeland was then Deputy Secretary (Policy and Planning), DOT and previously First Assistant Secretary, Air Transport Policy Division, DOT.

[18] Thus Freeland, *Hansard*, *SCT*, 17 August 1978, p. 5591 incorrectly claims that overcapacity forced up costs and hence prices in the U.S.; the direction of causation was the reverse.

of civil aviation. It notes an increase in average load factors which was generally greatest on the most competitive routes. Similarly, there is little evidence of overcapacity in the post-war decade of supposed chaos in the Australian industry. Goodrich (1960, p. 138) notes that during the period 1945-1952 the provision of excess capacity could not be listed as a reason for either instability or poor profitability. For the period 1952-1957 he also remarks:

> It has already been noted that there was virtually no control over aircraft importations . . . From time to time, concern had been expressed about excessive competition in re-equipment and consequent procurement of excess capacity . . . it is quite clear that the fears about the procurement of excess capacity were not justified by the facts.[19]

Airline re-equipment

Competition among airline operators is often alleged to hold harmful implications regarding the related issue of the acquisition of new equipment. Paltridge states:

> It is fundamental to the Two-Airline System that the airlines should not embark upon an equipment race as this would constitute a most damaging form of wasteful competition and would be contrary to the public interest.[20]

Goodrich (1960, p. 219) notes that aircraft equipment decisions had a major impact on the performance of the Australian airlines during the post-war decade. For five years Australian National Airways (ANA) made little effort to provide an adequate alternative to TAA's Convair 240 which became Australia's first pressurised aircraft when introduced in 1948. TAA again made a superior aircraft choice when in 1954 it introduced Australia's first prop-jet, the Vickers Viscount; ANA had decided to operate the piston-engined DC6Bs in the previous year. The consistent inability of ANA to provide aircraft which matched the

---

[19] Goodrich (1960, p. 175). Goodrich qualifies his statement by suggesting that delays on aircraft imports due to dollar shortages may also have exerted a restraining influence on capacity.

[20] Quoted in Brogden (1968, p. 158).

passenger appeal of its competitor was a significant determinant of its financial plight [see Brogden (1968, p. 78 and p. 103)]. Observing this Australian and some overseas experience, many concluded:

> Unfettered competition could result in an equipment 'rat race' beyond the capacity of the industry to handle economically. In a small nation such as ours the whole industry could be jeopardised. [Hillyar[21] (1969, p. 19)]

In 1958 the Government felt:

> . . . it was clear that the stage would be set for a struggle by each airline to out-equip the other, regardless of the capital cost involved.[22]

Its response was to reject the airlines' requests for new aircraft (Electras for the Ansett organisation and Caravelles for TAA), and to exert a greater influence on equipment choice through the Airlines Equipment Act. Since comparable equipment was regarded as a prerequisite for stability, the airlines were obliged

> . . . not to introduce aircraft of a type which, having regard to the types already in operation, would be deterimental to the stability of the air transport industry. This . . . obligation is, of course, designed to stop a wasteful re-equipment race leading to a multiplicity of new and expensive aircraft types.[23]

While Australia has not yet experienced any significant disruption through equipment choice under the TAP, it is not clear that an open airline market would lead to such 'rat races' or that the public has benefitted greatly from this feature of the policy. The Australian post-war experience

---

21 Then Secretary of Ansett Airlines of New South Wales.

22 Paltridge, *Hansard, S*, 1 October 1958, p. 757.

23 Paltridge, *Hansard, S*, 1 October 1958, p. 757. Paltridge also remarks that the two major operators willingly accepted this diminution of their managerial responsibilities. Note that this equipment race scenario seems to conflict with the argument, discussed in the previous section, that competition will hinder technological progress in aviation.

cannot be considered a rat race; rather it was the market penalising an operator, ANA, for making inappropriate choices of equipment. Thus any instability was confined to the relative market positions of the operators involved. In this regard the recent decisions of the two major operators to re-equip with wide-bodied aircraft of different capacities, the Airbus A300 and Boeing 767 for TAA and AAA respectively, is a potentially significant development for the industry. The use of different aircraft offers the possibility of increased competition between the operators through a greater differentiation of airline services, e.g. different schedules and frequencies, and perhaps even different fares if there are differences in operating costs and if the regulators permit such cost differences to be reflected in fares. If the use of different aircraft does have a significant impact on consumer choice it will be interesting to observe the Government's reaction this time to the self-induced market 'instability'.

The U.S. experience relating to re-equipment can be directly attributed to the poor system of CAB regulation [see Kahn (1971, p. 213)]. As noted above, the policy of controlling mainly prices stimulated non-price competition and hence placed greater emphasis on the type of aircraft in service. The problem was compounded by the failure of the U.S. regulatory authorities to allow quality differences in aircraft to be reflected in price differences; the CAB insisted that all operators charge a common fare for a given route service. Hence it became imperative to fly the most popular type of aircraft since there was no possibility of competing with a different price/aircraft type product mix. It follows that the responsibility for any wasteful underutilisation of older or less popular equipment lay with the regulators rather than open market forces.

Another theme of the re-equipment issue is the effect of depreciation on prices and hence consumer welfare. It is frequently argued that aircraft must be used for a reasonable time to allow adequate obsolescence and to reduce the frequency with which heavy capital outlays are needed.[24] For example, Goodrich (1960, p. 219) argues:

> . . . airlines' operating costs would be inflated because of the high obsolescence and depreciation

---

24 This view is expressed by Hillyar (1969, p. 19), Poulton (1959, p. 32), Richardson and Poulton (1968, p. 79) and Paltridge, *Hansard*, *S*, 1 October 1958, p. 757.

rates necessary when frequent re-equipment is forced upon them.

However, this argument has several shortcomings. Many industries are subject to the risk of technical obsolescence of their capital equipment. This is a naturally occurring risk which must be planned for in their investment decisions. The depreciation rates which are allowed for taxation or accounting purposes represent an expectation of the rate of decline in the economic value of an industry's capital assets. In an open market unanticipated technical obsolescence is reflected in a faster decline in the capital value of a firm's assets (assuming that there is no offsetting compensation in the taxation system). Consumers do not bear this burden but rather enjoy the benefits from the introduction of the new technology. In this environment managers have a clear incentive to make good investment decisions with respect to the technology of their operations. Managers who consistently make poor equipment decisions are likely to be replaced through takeover or bankruptcy, e.g. ANA's experience in 1957. Consumers suffer from technical obsolescence only in a regulated environment where policy aims to maintain the financial position of the existing firm. They are then forced either to continue to be served by inferior equipment[25] or to pay higher prices which cover the additional capital losses which have occurred because of an inappropriate choice of equipment.

The argument regarding depreciation and technical obsolescence thus reduces largely to one of shifting the risks and costs of such events from shareholders of current firms to consumers. Paltridge confirms that Australia's aviation policy was so designed:

> Unless the airlines are permitted to sustain disastrous capital losses by replacement of existing equipment before it is adequately obsolesced, we must plan on the present front-line equipment remaining in operation for at least several more years.[26]

---

25 Brogden (1968, p. 137) reports that in March 1957 the Government rejected an application from TAA to introduce Caravelle jet aircraft into service in 1959. The first pure-jet service was subsequently operated in November 1964 with Boeing 727s.

26 *Hansard*, *S*, 28 September 1961, p. 722.

Parallel scheduling

Another example of wasteful competition caused by the pattern of regulation adopted by air transport authorities is Australia's controversial parallel scheduling. The practice is well documented and pervasive [see Gannon (1979a)]. Its waste lies in the restriction of choice of flight times and the extra demands placed upon airport infrastructure. Most observers agree with the proposition that parallel schedules are the natural result of competition in a system in which all the major attributes of the operators' products are identical, e.g. fares, aircraft type, total capacity and in-flight services. In these circumstances timetabling is the most significant means of non-price competition available. With this background, two operators and no threat of entry, simple economic theory suggests a tendency towards parallel schedules.[27] However, the theory also suggests that this tendency would weaken if market entry were possible, if there were not only two operators, or if alternative means of competition, especially with prices, were possible. In this regard the recent choice by TAA and ATI of new wide-bodied aircraft with different capacities can also be expected to lessen the incidence of parallel schedules. However, the important point is that once again the fundamental cause of the observed wasteful competition is the existing regulatory policy.

Summary

This section indicates that the major forms of wasteful competition observed in airline markets are directly attributable to government intervention in the operation of open market forces rather than to these forces themselves and are unlikely to persist in a free market environment. There are two basic strategies available with which to respond to such wasteful competition: attempt to cast the regulatory net still further, or move to open competition and rely on market forces to satisfy consumer demands.

---

[27] See Gannon (1979a) and Hocking (1972, Ch. 3). These results are based on the principles of spatial competition, of which Hotelling (1929) is the seminal work.

III. NETWORK CONSIDERATIONS

Allegations of abandoned services

An increased likelihood of deterioration or destruction of airline networks is another undesirable consequence frequently alleged to follow the introduction of open competition in airline markets. Teague (then Manager, Planning and Development, East-West Airlines [EWA]) states:

> If in fact there is total deregulation I would be very very apprehensive about the possible results . . . But in total I could see large areas of the country or large numbers of our communities receiving very very much worse services than is the case today.[28]

Similarly, AAA (1979, p. 15) argues:

> . . . a predator third operator operating only on the main trunk routes would affect the existing airlines in their important role in providing high quality services to lesser developed areas.

In analysing these claims it is useful to consider two possible regulatory situations. Firstly, there is the case where costs are covered on each route but profitability varies among routes. Secondly, there may exist cross-subsidisation where the losses on some routes are offset by the profits on others.

Routes of varying profitability

In the first case it is alleged that if the industry is deregulated existing and new firms will concentrate their activities in the most profitable markets to the neglect of those routes which are less profitable. The possibility of curtailment of such services underlies the unwillingness of the Australian Federation of Consumer Organisations [AFCO] (1978) to advocate complete removal of entry restrictions into airline markets. Ansett (1965, p. 13) appears to hold a similar view:

> Additional operators interested only in the few routes with high density end-to-end traffic could impair the

[28] *Hansard*, *SCT*, 16 May 1978, p. 4910.

overall competitive nature of the trunk route networks.

In this case any shift in capacity towards the most profitable routes would be due to the current restrictive regulation which is preventing operators from pursuing profitable investment opportunities. Wholesale market abandonments are unlikely since these services are profitable; although additional investments of existing and new operators will naturally be attracted in the first instance to the more lucrative routes. In these circumstances it is most likely that any difficulties will reflect short run adjustment to the enlarged set of investment opportunities, and in the longer term one could expect the maintenance and possible increase in services on all the profitable routes.

Cross-subsidisation of airline networks

Competition might be considered potentially more disruptive when cross-subsidisation exists within airline networks. The cross-subsidisation objective was warmly embraced in the past and at times appears to have significantly influenced policy. Brogden (1968, p. 42) and Goodrich (1960, p. 62) both report that the Corbett Committee[29] considered that a large airline should be compelled to operate services on developmental routes, offsetting any losses from the profits earned on denser routes. One perceived advantage was the resultant reduction in the level of required government subsidies. The notion of cross-subsidisation was apparently well received at that time and used by the Australian Labor Party (ALP) as an argument in support of its attempted nationalisation of the airline industry in 1945.

The attractiveness of cross-subsidisation was not confined to only one political party; Menzies (then Leader of the Opposition and subsequently Prime Minister, 1949-1966) also saw the need for profitable routes to balance the losses on developmental routes.[30] In later years it became an accepted part of the TAP. Consideration of this objective was an important factor in the selection of ATI as the new

---

29 This was an Interdepartmental Committee on Civil Aviation with A.B. Corbett (first Director-General of Civil Aviation, 1939-1944) as Chairman. Its report was submitted in December 1943; the major findings, while not made public, were apparently well known.

30 *Hansard*, *HR*, 25 July 1945, p. 4555.

private partner in the policy in 1957:

> . . . only Ansett Transport Industries Limited was prepared to proceed on the basis of the Government's declared policy of subsidising unprofitable feeder services . . .[31]

Cross-subsidisation continues to have support; Smith (1978, p. 6) reports a statement by Nixon (Minister for Transport, 1975-1979) indicating that he considered the Sydney/Melbourne route to be a legitimate source of finance for the less profitable routes.

Cross-subsidisation could be expected to substantially disappear if free entry into the industry were allowed.[32] Thus the maintenance of internal subsidies requires a system of economic regulation which prevents entry into the profitable markets. Not surprisingly, the existence of cross-subsidisation is often used to support a continuation of the current restricted entry under the TAP. For example, as a consequence of breaking down this policy, Chippindall (1965) foresees:

> A tendency for the new operators to confine their activities to the more profitable routes, leaving the two main operators to bear the burden of the mass of unprofitable routes whose servicing is part of the overall responsibility for providing airline services which they have undertaken.

When assessing the desirability of cross-subsidisation the various arguments under consideration can be grouped into two broad classes: efficiency and equity. Economists are generally quick to reject cross-subsidisation on efficiency grounds. Since prices are not permitted to reflect marginal costs, resources are misallocated (there is a relative undersupply of air services in profitable markets) and an efficiency loss is imposed on the community.

---

31 Paltridge, *Hansard*, *S*, 13 November 1957, p. 1211. ATI accepted this cross-subsidisation obligation in return for a strengthening of the Rationalisation procedures and the provision of loan guarantees.

32 As will be argued shortly, the disappearance of cross-subsidisation is not necessarily equivalent to the disappearance of air services on the currently unprofitable routes.

It may be argued that government intervention is desirable; competition may be imperfect or externalities may exist. Some issues of relevance to this argument are discussed in greater depth in Chapter 5. Yet even if such claims are valid, economists seem virtually unanimous in advocating the use of direct subsidies to deal with such problems.[33] The fact that government expenditures would be higher under such an approach is a misleading consideration since society incurs the real opportunity costs of the relevant services no matter how they are financed. It might also be argued that greater electoral opposition may arise to government initiatives involving direct subsidies; hence a system of cross-subsidisation may be the most politically effective means of attaining the desired objective. This argument suggests that it is the policy instrument itself rather than the policy goal per se which the community finds objectionable. However, the observed difference in community reaction is most probably due to a relative lack of information on the costs and benefits of cross-subsidisation rather than any intrinsic preference for this policy technique. One might hold doubts on the appropriateness in a democratic society of any policy which is incapable of being adopted when subjected to the public scrutiny of the government budgetary processes.

Even the equity arguments for cross-subsidisation are unconvincing. The equity justification seems to be based on the notion that air services should be available to all members of the community and that all consumers should pay an equal fare for what appears at face value to be the same product, i.e. equal distance travelled. Kahn (1971, p. 244) summarises an alternative viewpoint:

> . . . advocates of entry restrictions in the trunk airline business have justified the internal subsidisation that they protect on the ground that small towns 'deserve air service' as much as large. It is not clear who, exactly, the people are in those towns who 'deserve' such service: presumably, they are the relatively well-to-do who can afford travel by air; nor is it immediately obvious by what morality

---

33 E.g. Levine (1965, p. 1428), Keplinger (1976, p. 198) and Smith (1978, p. 6). Payment of direct subsidies was the most important instrument of government influence in the early years of Australian civil aviation [see Goodrich (1960)].

> they deserve to be subsidised in receiving such service, either by general taxpayers or by regular travellers over the routes . . . that can generate enough traffic to pay their way.

Cross-subsidisation clearly has important distributional implications. However, it seems preferable to base any redistribution policies on income or wealth rather than area of travel or residence. Perhaps it might be argued that the operation of certain unprofitable air services is to the net benefit of the nation. If this is so, why should only travellers on more profitable routes bear the cost of providing these services? These factors suggest to many that the cross-subsidisation of airline services can often be inequitable.[34]

The previous quote by Chippindall may be taken to imply that it is unfair that existing operators bear the burden of unprofitable routes when entry is unrestricted. However, this interpretation is of little value since any call for freedom of entry also, most sensibly, involves freedom of exit. Another strange perception of equity is presented by Confederation of Australian Industry (1980) which argues that cross-subsidisation in the airline system is acceptable and in the national interest since it would be 'unreasonable' to expect the major operators to set fares which ensure that all routes cover their costs!

The policy of cross-subsidisation appears to have fallen from favour in recent years and is increasingly rejected. Transport regulators often express concern about the existence of cross-subsidisation[35] and argue that pricing must reflect 'true costs'.[36] The fare recommendations of DOT (1979a, p. 80) involve a closer alignment between prices

---

[34] E.g. Levine (1965, p. 1428), Kirby (1979, p. 110) and Rushton (Western Australian Minister for Transport), *Australian Financial Review*, 7 February 1980.

[35] Rowland, *Hansard*, *SCT*, 10 May 1978, p. 4713. Rowland was then First Assistant Secretary, International Policy Division, DOT.

[36] DOT, *Annual Report*, 1976/77, p. 18. While this statement most probably refers to networks as a whole, it is but a short logical step to argue that route prices should reflect route costs. However, DOT (1980b, p. 24) is not prepared to go this far. It recommends the continuation of fare setting using a 'nationally consistent approach' on the grounds of public acceptability and administrative convenience.

and costs. This aim was further reflected in subsequent adjustments to the airlines' fare formula which made more accurate allowance for economies of flight length and which changed the ratio of first class to economy class fares. Nixon also found it difficult to tolerate unlimited cross-subsidisation. He supported TAA's unsuccessful attempt to sub-contract two unprofitable Queensland routes, Brisbane/ Mt. Isa (via ports) and Townsville/Mt. Isa (via ports), to Bush Pilot Airways (BPA) and suggested, not too subtly, that he would consider directing the maintenance of the existing TAA F27 services if 'the State is prepared to subsidise the losses incurred on these services' [DOT (1979c)]. Finally, the airlines themselves express some opposition to cross-subsidisation:

> The industry cannot accept the proposition that air travellers on major trunk routes . . . should subsidise travellers on rural or outback services from which the Government has no chance of recouping its expenditures. [Ansett (1976, p. 54)]

However, in spite of this statement, it was noted above by Paltridge [see footnote 31] that Ansett found cross-subsidisation to be acceptable when combined with other more favourable policy features.

Effects of competition on airline networks

While the above discussion examines the general issue of cross-subsidisation, an assessment is still required of the claim that many services would be lost if the present policy were abandoned. The potential for disruption might be expected to depend on the extent of cross-subsidisation now present in the airline system. Although no detailed cost information is currently available,[37] the evidence suggests its existence. Hocking (1972, p. 38) concluded that the industry was characterised at that time by cross-subsidisation. The fare formula which was subsequently introduced in August 1974, and which was based on the average costs of servicing an entire network and consisted of a flagfall (reflecting the cost of terminal facilities) and a rate per kilometre (covering

---

[37] One of the main tasks of the recent Independent Public Inquiry into Domestic Air Fares is to report on the degree of cross-subsidisation in the networks of the two major operators.

operating costs), was a recipe for cross-subsidisation since it failed to recognise that passenger costs can vary across routes. In particular, the fixed cost component varies with traffic density and operating costs per kilometre decline as stage length increases. If regulation has been successful in limiting total airline profits to normal levels, the fare formula would seem to imply cross-subsidisation, not merely price discrimination. The investigation by Gannon (1979b, p. 147) confirmed that the air fare formula then resulted in an apparent cross-subsidisation of short distance routes by long distance routes. Gannon also concluded that the F27 network as a whole was cross-subsidised from other services. Further evidence on the existence of cross-subsidisation is provided by the airlines themselves who claim to operate some unprofitable services. Forsyth (1979, p. 69) notes reports indicating that in 1976/77, when its profit was about $7m, TAA lost $1.5m on the two Queensland routes mentioned above. Nixon confirmed that TAA was unable to cover its operating costs on these routes [DOT (1979c)].

Even though substantial cross-subsidisation may exist, it does not necessarily follow that open market competition and the resultant tendency to eliminate such internal subsidies will cause a massive loss of services throughout the country. While an existing airline may be unable to achieve profitable operations on a certain route, it is not valid to conclude that some other carrier (with different schedules, equipment, etc.) could not do so. In an open market other operators would most likely be quick to service any route vacated by the two major airlines. This contention is supported by the U.S. deregulation experience where few routes have been abandoned and not taken over by others [see ITA (1979)]. In addition, Australian commuter operators have generally taken over unprofitable feeder services which the major carriers have ceased to operate. It is quite possible that many previous loss routes will enjoy increased services from their new operators.[38]

Some may claim that the crucial issue relating to airline network considerations is not the argument regarding service versus no service, but rather airline services versus commuter operator services. Thus, even though the replacement commuter operator may provide a more frequent service,

---

[38] See Bureau of Transport Economics [BTE] (1978, p. 58). BPA (1977, p. 7 and Appendix 2) gives numerous examples of routes which it has taken over from the major operators.

travellers may still view this as inferior to the previous operations of a major airline. An example of this was the community hostility towards TAA's proposal to sub-contract some Queensland F27 services to BPA. It is doubtful that this preference would be so strong if the local community involved were obliged to pay the full cost of the alternative services.

Summary

The discussion in this section indicates that little effort has been made to provide a valid justification for the cross-subsidisation apparently now present in the airline system. Apart from the aim to reduce government expenditures, it appears that any public welfare and income redistribution goals which are supposed to underlie current policy can be more effectively achieved by direct subsidies. If this proposition is accepted, it follows that the true impact of the policy of economic regulation of airline markets is not to protect air service networks, but rather to achieve income transfers between various consumer groups in a politically effective manner and to provide a rationale for shielding existing operators from the effects of competition.

IV. CONCLUDING REMARKS

When discussing civil aviation deregulation in the U.S., Hewitt (Chairman of Qantas, 1975-1980) states:

> This is not an industry in which cut-throat competition is good for the consumer.[39]

However, this chapter suggests that there is little evidence to support the assertion. Apparent examples of the undesirable effects of competition usually can be more accurately traced to poor regulation rather than open market forces per se. It appears that the inefficient operator, not the consumer, has most to fear from competitive airline markets.

---

39 Quoted in *Traveltrade*, 21 January 1980.

2
Competition Among Airlines: Too Little?

Fears of monopoly

Fears that the Australian domestic air transport industry would develop into a monopoly and that this would be inimical to the community's welfare have always loomed large in the airline regulation debate. Drakeford (ALP Minister for Civil Aviation, 1941-1949) claimed

> . . . abundant evidence that a monopoly is inevitable in the near future.[40]

Dedman, then ALP Minister for Post War Reconstruction, viewed the threat as more immediate:

> . . . the Government is entirely opposed to private monopolies . . . in my opinion, the present airline operators in Australia have a monopoly of that industry . . . the conduct of a monopoly by private enterprise is, in effect, immoral.[41]

Responding to this perceived threat of monopoly:

> The Chifley Government decided that something should be done to prevent that monopoly [Australian National Airways Proprietary Limited] from fleecing the Australian people.[42]

Antipathy towards an air monopoly was not limited to the one political party. Menzies stated that it was

> . . . no part of the policy of the Government to foster

---

40 *Hansard, HR*, 18 July 1945, p. 4179.

41 *Hansard, HR*, 25 July 1945, p. 4559. A crucial word in this statement is 'private'.

42 Drakeford, *Hansard, HR*, 30 October 1952, p. 3967.

either a government monopoly or a private monopoly on the major air routes.[43]

Even in more recent times the fear of monopoly continues to influence air transport policy considerations. At a recent Federal Council of the Liberal Party, a Western Australian resolution recommended the sale of TAA,

> . . . with appropriate legislative provisions designed to prevent a single monopoly taking control of the major national airline system.[44]

The natural monopoly argument

The theoretical economic basis underlying this attitude towards a monopoly market structure and calls for government intervention is the so-called 'natural monopoly' argument. If the technological conditions of production are such that the firm's long run average costs decline over the entire market size (i.e. there are economies of scale), then a single firm will always be able to supply any given quantity of output at less cost than if there were more than one firm. Hence there is a natural tendency for the industry to become dominated by a single firm. The argument asserts that, once this situation is established, the single profit maximising firm will be able to exert its monopoly power and earn supernormal profits.

This simple account of monopoly pricing associated with economies of scale can be represented in the following diagram.

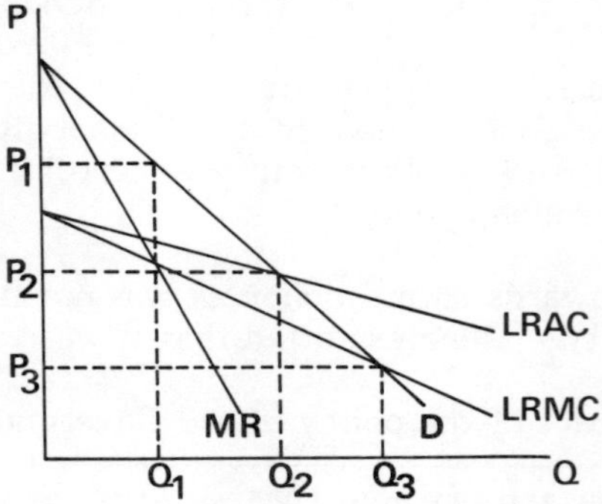

[43] *Hansard, HR*, 21 November 1951, p. 2399.
[44] Reported in *Daily Telegraph*, 9 April 1980.

A firm with monopoly power maximises its profit by producing where marginal revenue equals marginal cost; then (P_1, Q_1) is the monopoly price/quantity outcome. Advocates of the natural monopoly argument prescribe regulation to prevent entry (so that a single firm can take advantage of the economies of scale) and the control of price. Price may be regulated to the zero profit level P_2, or ideally (to achieve Pareto optimality) to P_3, where some arrangement can be made regarding the resultant losses of the firm. Note that already there is an apparent inconsistency in the policy prescription: the desire for regulation is motivated on the one hand to capture the benefits of economies of scale (which are allegedly endangered if firms are free to enter the industry) and, on the other hand, to prevent monopoly pricing (which supposedly occurs if there is only one firm).

This simple theoretical argument has frequently been presented and discussed in the airline industry context.[45] Two questions must be investigated. Firstly, is the natural monopoly argument applicable to the Australian air transport industry, i.e. are there economies of scale in the provision of airline services in this country? Secondly, is the argument valid from a theoretical viewpoint?

Economies of scale

In the course of the Australian airline regulation debate many of the participants have claimed the existence of significant scale economies and hence a natural tendency towards monopoly. Brogden (1968, p. 60) reports that during the post-war decade all except Ansett believed that there was room for only one major airline on domestic routes. Drakeford considered ANA an 'inherent monopoly'.[46] The Corbett Committee concluded that it was desirable to have only one airline company operating all Australian air services, since it would be able to do so more efficiently and economically than a larger number [see Goodrich (1960, p. 61)]. This conclusion was also supported by the Nixon review of 1951.[47]

---

[45] E.g. Levine (1965, p. 1424), Keplinger (1976, p. 195), BTE (1978, p. 29), Smith (1978, p. 10) and Grenning and Coat (1979, p. 10).

[46] *Hansard, HR*, 30 October 1952, p. 3975.

[47] E.V. Nixon and Partners, a firm of chartered accountants, was appointed by the Menzies Government to investigate the economic and financial basis of the airline industry [see Brogden (1968, p. 94)].

Economies of scale appear to have played a major role in the decision of the Menzies Government to introduce its TAP. In a letter to the Chairman of ANA dated 28 March 1952, Menzies wrote:

> . . . we have discussed the future of the airlines in terms of two operators because . . . in our present state of development there is no room for more than two major national operators if the necessary standards of efficiency are to be maintained.[48]

Again, at the time of the strengthening of the TAP in 1957:

> . . . [the Government] could not escape the definite conclusion that even the trunk route system could not support more than two major airlines.[49]

Some academic contributions have also given support to the claim that scale economies exist in the airline industry. Goodrich (1960, p. 149) argues that if the industry consisted of only private firms then a monopoly would probably develop as a result of economies of scale.[50] Similarly, Hocking (1972, p. 50) concludes that there were still considerable economies of scale to be reaped in the Australian market; if firm size were doubled, unit costs might fall by 10-20 per cent.

More recently DOT (1979a, p. 74) found:

> the size of the Australian market would not support a third carrier on the trunk network as a whole and therefore the Government should continue to control entry . . . to maintain the Two Airline Policy . . . ,

while Dignam (1979, p.10), then Assistant General manager of TAA, argued:

> Our small domestic market means that optimum

---

48 Quoted in Goodrich (1960, p. 142).

49 Paltridge, *Hansard, S,* 3 September 1957, p. 82.

50 Goodrich's final position on this question is unclear: he also recognises that a private monopoly is not inevitable (p. 149), argues that a TAA monopoly would not likely have developed if ANA had left the industry (p. 140), and concludes that the conditions for the natural monopoly argument do not apply to air services (p. 84).

economies of scale are not always achieved.

Another popular viewpoint is that a private enterprise monopoly is inevitable if the government-owned TAA should cease to operate.[51]

However, a careful consideration of the nature of airline costs raises doubts as to the applicability of the natural monopoly argument to this industry. The prerequisite for economies of scale is declining long run average costs. On the demand side economies of scale may arise from the number and diversity of markets served. Variations in demand may tend to compensate each other causing economies of operation. However, as Kahn (1971, p. 123) notes, diversity of demand is not necessarily an argument for monopoly; it is more likely to indicate the desirability of integration of operations across markets.

Economies of scale more typically are associated with the presence of substantial fixed costs in an industry. It is not however the mere existence of fixed costs which indicates economies of scale; this point is obvious when it is remembered that in the long run all costs, by definition, are variable. The relevant issue is whether or not there are increasing returns from the provision and utilisation of larger amounts of these fixed cost facilities. Thus statements pointing out the degree of fixed costs associated with airline operations do not, by themselves, provide conclusive evidence regarding economies of scale. Such statements are more likely to provide information on short run average costs and hence are of more direct relevance to the question of destructive pricing which is examined in Chapter 1.

While the presence of significant fixed costs does not necessarily prove the existence of economies of scale, statements regarding the absence of such costs can often be more informative on the issue. Since the main potential source of economies of scale lies in the provision of fixed cost capital assets, a low ratio of fixed to variable costs probably lessens the chance of such scale effects. Hence the strong consensus, noted in the previous chapter, among economists that the airline industry is characterised by a low ratio of fixed costs to variable costs supports the contention that substantial economies of scale are unlikely.

The presence of economies of scale in the airline

---

51 See Brogden (1980, p. 117), Morris (1980b) and TAA (1980). Morris is the ALP Shadow Minister for Transport.

industry has also been examined in a wide range of empirical studies. This research indicates that, while network characteristics such as average stage length and traffic density are important determinants of airline costs, firm size per se has little independent influence on average costs. The common conclusion is that economies of scale are not significant beyond a certain minimum efficient scale of operations of around five aircraft.[52] Mackay (1979) estimated a regression model of airline costs and found that if TAA and the Ansett airlines were merged then average costs would decline by 4 per cent. Alternatively, if a third airline were introduced and each airline had equal market shares, then average costs would rise by 2.5 per cent.[53] In summary, there is a lack of statistical evidence supporting the existence of substantial economies of scale in Australian airline operations.

Theoretical validity of the natural monopoly argument

While most researchers have rejected the applicability of the natural monopoly argument to the airline industry, there has been no serious questioning of the argument's theoretical validity. For example, Levine (1965, p. 1424) claims that there is 'general agreement upon the economic necessity of regulating natural monopoly industries' and Smith (1978, p.12) considers it apparently 'unimpeachable'. However, Demsetz (1968) drew attention to a non sequitur in the natural monopoly argument. Although economies of scale may result in a single firm in an industry, it does not necessarily follow that this firm has the ability to monopoly price. That ability comes from establishing barriers to entry into the industry and these two phenomena, scale economies and entry

---

52 See Grenning and Coat (1979) and Smith (1978) for a review of these issues. In June 1979 the fleets of TAA and AAA each included 10 B727-200s and 12 DC9s [DOT, *Annual Report*, 1978/79, p. 162]. In his public evidence to the Independent Public Inquiry into Domestic Air Fares on 25 November 1980, Teague supported the view that scale economies were negligible beyond the minimum fleet size.

53 These results implicitly assume that changing the number of operators will not signficantly affect other determinants of airline costs, e.g. average stage length and route density.

barriers, are quite distinct.

The above diagram illustrates the argument. The natural monopoly outcome (P_1,Q_1) is not a stable equilibrium. The supernormal profits earned by the firm at that point attract other firms who, given no entry barriers, can enter the industry, undercut the monopolist and capture his entire market. Thus there are pressures for the price to fall below P_1. In the simple exposition of the natural monopoly argument which is usually presented the only stable equilibrium appears to be the zero profit outcome at (P_2,Q_2). McKenzie and Tullock (1978, p. 217) even show the possibility of the natural monopolist producing the Pareto efficient output, by selling Q_2 units at price P_2, then expanding output to Q_3 through price discrimination.

This analysis implicitly assumes that firms can costlessly enter and leave the industry. In this simple world of zero adjustment costs there can be no deterrent to entry into the industry since if entry is unsuccessful the firm is able to shift its capital resources into other activities without penalty. In this setting the duplication of capital assets also imposes no social cost on the community.

In reality adjustment costs do occur. A major determinant of these adjustment costs is the extent to which the unsuccessful firm can recover its capital investment, i.e. the value of its assets in alternative uses. Costs of entry and exit will be relatively high when a firm's production process is characterised by significant sunk costs resulting in its capital assets having little value in alternative uses. This situation most often occurs when capital assets are long lasting and highly specific. The presence of these adjustment costs opens the possibility of deterring entry since the existing firm is able to ignore its sunk costs in any potential competitive pricing battle. Likely outcomes in this situation depend on the assumptions made about firms' reactions to the initiatives of competitors; these cases can only be investigated with more complex analysis. However, the point remains that it is the nature of capital assets, i.e. the importance of sunk costs, which influences a firm's adjustment costs and hence ease of entry, and this consideration is independent of the question of economies of scale.

This discussion can be illustrated with some simple examples. A railway track between two country centres probably has little value in alternative uses, e.g. digging up and laying elsewhere or using as scrap. Hence a significant part of the value of the track will represent sunk costs. This feature is likely to deter entry, give railway firms some power to price above average costs, and impose social costs on the community if the track facilities are duplicated by competitors. However, such is not the case for airline operators. The previous chapter noted that their fixed cost assets seem quite mobile and possess a relatively high value in alternative uses. Unsuccessful airline firms are thus able to recoup a significant proportion of their capital investment. In these circumstances it is likely that entry will be easier, airlines will have less pricing power and social costs from duplication will be minimal.

The threat of competition

A common error in discussions of market structure is the failure to distinguish between the concepts of competition and market concentration. Thus attempts to increase the competitive nature of an industry by fixing the number of firms operating in the industry (e.g. a Two or Three-Airline Policy) have no guarantee of success. Similarly, when entry and exit costs are small a high degree of market concentration holds little threat to consumer welfare. The influence of the above ideas is reflected in the importance given by some recent writers to the market policing roles of 'potential competition' and the threat of entry.[54]

ATI (1980, p. 19) claims that it is impractical to suggest that the threat of competition would have a beneficial effect. Its argument is based on the long lead time which it alleges is necessary to put new aircraft equipment into operation. Yet it seems clear that there are many potential sources of competition for the major domestic operators, e.g. international and regional airlines and commuter and freight operators. Market access and the availability of aircraft through import controls appear to be more significant barriers than lead times. For example, Qantas (1980) claims that it could immediately operate long-haul domestic services with economy fares of about 70 per cent of those currently charged. Cabotage, the use of the spare capacity of international carriers on domestic routes, also offers prospects of immediate different choices of air services.

Kahn (1971, p. 12) summarises this discussion:

> The possibility of competitive entry is the principal

---

54 E.g. BTE (1978, p. 60), Forsyth and Hocking (1978, p. 4) and Kirby (1979, p. 113).

> limitation on monopoly power in a market economy.[55]

Also in this regard Walsh (1951, p. 44), then Acting General Manager of ANA, clearly recognised, and later felt, the real limits to his firm's supposed monopoly power:

> ANA would have no monopoly except the monopoly that arose from doing the job better than the others could. For private enterprise can compete with us any time it likes. If we fell down it would. Such competition would be fair competition. We both would not survive - that would be an economic impossibility. The most efficient would survive. And that is one of the major virtues of freedom of enterprise.

Barriers to entry

The above discussion leads to the conclusion that the existence of economies of scale is not a sufficient condition for the monopoly pricing outcome to result and hence for a justification of government intervention. We must consider whether or not there are significant barriers to entry into the airline industry.

Callison (1975, p. 776) considers the possibility of new entry into air transport 'naive'. Effective entry barriers are frequently alleged to come from the high capital costs of the industry, product differentiation and absolute cost advantages of existing operators. Hocking (1972, p. 49) suggests that these provide 'significant economic obstacles' to potential entrants.

While the current operations of the major airlines undoubtedly represent a large business undertaking and modern aircraft equipment is not inexpensive in absolute terms, this does not necessarily exclude entry. Apart from the sunk costs issue examined above, this argument must rely on capital market imperfections which bias the availability of funds for large scale investments. However, while the size of the necessary investment has increased over the years, the capital market in Australia has also become more developed,

---

[55] Kahn also notes:

> No barrier to entry is more absolute than one imposed and enforced by the sovereign power of the state.

making it difficult to draw firm conclusions as to whether it would be harder to finance a new airline operation now than in the past. It is also often assumed that new entrants would be of a similar size to the current major operators. DOT (1979a, p. 74) adopted this view when it rejected competition on the trunk routes. However, this presumption is not necessarily the case. A new operator could start small by concentrating on only part of the entire network and, if successful, grow over time.

It is unlikely that product differentiation forms a major entry barrier. The nature of the product is such that any firm offering a price/service combination more in tune with consumer preferences will probably find ready market acceptance. Similarly, since airline technology on both the marketing and equipment sides is available to all, it is considered unlikely that there are significant absolute cost advantages in favour of the existing operators.

Additional evidence on the question of whether there will be too little competition in an open market for air services comes from the empirical observation of such markets. The Californian intrastate market revealed no tendency for a monopoly to develop and the success of the intrastate Pacific Southwest Airlines demonstrates the ability of an efficient, specialist operator to survive [see Levine (1965, p. 1439)]. The Australian experience also provides some evidence on the relative ease of entry into airline markets and the improbability of monopoly pricing behaviour. Goodrich (1960, p. 89) notes that, despite the highly concentrated nature of the industry (ANA held 80-90 per cent of the market), there was, after allowing for wartime conditions, no evidence that ANA either achieved monopoly profits or adopted restrictive practices in establishing and maintaining its market position in the period to the end of the Second World War. During the post-war decade, a relatively competitive period for Australian aviation, Ansett was able to increase its market share by specialising in the tourist market and offering a lower price/lower quality choice. Similarly, TAA offered a service that was so readily accepted that it captured over 40 per cent of the market on inter-capital routes within a year of commencing operations in 1946 [Goodrich (1960, p. 97)]. TAA's spectacular success demonstrates the ease of entry into airline markets when a new firm has a superior or innovative product to offer consumers, although it must be acknowledged that in this case the attractiveness of the new service was enhanced by the preferential treatment which

TAA received from the ALP Government of that time [see Chapter 3]. More recently, the widespread granting of commuter authorisations over the network of Ansett Airlines of South Australia and the resultant adverse effect on that regional airline again illustrates the ability of new operators to enter a market and to succeed in attracting consumer patronage away from existing firms.[56]

Summary

Monopoly exploitation of consumers has been one of the most popular themes of the Australian airline regulation debate. Yet the validity of this argument is most doubtful. There is virtually no empirical evidence to support the view that there are significant economies of scale in the provision of airline services. An examination of the nature of airline operations also suggests that, apart from government actions, there are few substantial barriers to entry into markets for air transport. Thus existing operators can be expected to have relatively little pricing power under open market conditions. Finally, the limited experience with competitive airline markets in the U.S. and Australia does not support the view that consumers will face monopoly exploitation in an airline industry which is free of restrictive economic regulation.

---

[56] See ATI, *Annual Report*, 1978/79. ATI questioned whether the policy of permitting such market entry by commuters was 'in accord with the best interests of the travelling public'.

3
A Question of Ownership

One of the special characteristics of the Australian domestic air industry is the presence of a government-owned firm operating in the same market as a privately-owned one. Hence it is necessary to examine the rationale for, and implications of, government ownership in the civil aviation field.[57]

The rationale for government ownership

The rationale for the introduction of a government-owned firm into an industry generally lies in the perception of an inadequate performance by private enterprise. Because private firms are influenced by the profit motive, it is sometimes believed they would not serve the national interest and might exploit the community. Thus, it is alleged, a government airline would offer superior service and safety to passengers[58] and prevent exploitation of the community by

---

[57] In addition to the state versus private enterprise debate, considerations of foreign ownership have also played a role in domestic civil aviation policy. Corbett (1965, p. 49) and Brogden (1968, p. 48) both suggest that a traditional and instinctive fear of British shipping interests was an influential factor in the ALP's aviation policy of the 1940s. It seems that Britain was not the only external threat; Calwell (Leader of ALP Opposition, 1960-1965) suspected that Ansett had 'a lot of American money behind him' [*Hansard*, *HR*, 30 September 1958, p. 1965]. Apparently there was no justification for discriminating among nations; the Liberal Party subsequently adopted the policy that there would be no foreign takeover or control of ATI [Cotton, *Hansard*, *S*, 13 April 1972, p. 1065].

[58] E.g. McKenna, *Hansard*, *S*, 1 October 1958, p. 765. McKenna was Opposition Leader in the Senate, 1951-1966.

private companies.[59] Often the desirability of a government firm is linked with the likelihood of the development of an industry monopoly. For example, Dedman asserts:

> . . . when an industry develops to the stage at which naturally a monopoly can give a more efficient service than can be given by a number of independent companies it should pass from private ownership to control by the Government. That is when it should be nationalised.[60]

In addition to the benefits of a government firm in its own right, it is sometimes argued that even greater advantages flow from a system in which government and private firms actively compete in the same market.[61] This argument is presented by Kahn (1971, p. 104), while Corbett (1965) is the leading proponent with respect to the Australian airlines. Corbett (1965, p. 117) introduces the concept of 'metaphism' ('the system of regulated competition between public and private enterprises, each having a substantial share of the industry's market') and claims that 'metaphytic competition' produces substantial political, administrative and economic benefits. Firstly, rivalry between the airlines provides the stimulus to draw out the best qualities of each system (the initiative of private firms and the public interest considerations of state-owned companies) and to suppress their respective undesirable qualities (exploitation and bureaucratic smugness). Secondly, each firm provides a ready yardstick with which to gauge the performance of the other. Closely related to this is the claim that the existence of a government firm is an aid in the economic regulation of the industry.[62] Finally, this system is alleged to provide political peace by satisfying the champions of both private and government enterprise.

---

59 E.g. McManus, *Hansard*, *S*, 20 November 1957, p. 1316.

60 *Hansard*, *HR*, 25 July 1945, p. 4563. This view is also supported in *The Age*, Editorial, 31 October 1961.

61 Hurford, *Hansard*, *HR*, 28 March 1972, p. 1257 offers a novel proposal to capture the benefits of both competition and state ownership: he suggests competition between two government-owned airlines!

62 DOT (1979a, p. 61) supports the retention of TAA on these grounds.

The experience from Australian aviation

Other observers of the performance of the airline industry under the TAP have concluded that the benefits of metaphism are, at least in this case, largely illusory. Instead of stimulating each other to greater heights, the two operators, free from any outside threat to their markets, may in fact settle for the 'quiet life'. So, while they may be observed to be performing equally, it may be equally poorly.[63] Wilson (1962, p. 41) notes that the yardstick benefit is limited in any practical sense; this is especially so for members of the public who do not have access to the data necessary to directly compare the performance of TAA and AAA. Finally, it is unlikely that the policy would placate the passions of those actively concerned with the state and private ownership dispute. The bitterness of past parliamentary debates concerning civil aviation provides testimony to this point.

One sideline to the ownership question is the controversy as to whether or not the state firm is less efficient than the private firm. Ownership per se has become regarded as increasingly irrelevant in determining a firm's performance; it is the aims, quality and techniques of management that are often considered to be more important.[64]

Certainly the ALP considered TAA to be significantly different from a private firm:

> . . . the primary aim of the Commission is not to seek profit, but to administer the airline with safety, efficiency and economy.[65]

However, one of the consistent thrusts of the TAP over the years has been to minimise the differences between TAA and its private competitor [see Appendix 1]. As well as the most obvious areas such as fleet comparability, capacity determination and the Rationalisation procedures, amendments to the Australian National Airlines Act have attempted to put TAA on a similar commercial basis to ATI. For example, the 1961 amendments set dividend targets for TAA and prevented

---

63 The economic performance of the major airlines is discussed in Chapter 6.

64 In the airline context see Wettenhall (1966, p. 72), Corbett (1965, p. 39) and Smith (1978, p. 105).

65 Evatt, *Hansard*, *HR*, 29 October 1952, p. 3874. H.V. Evatt was Leader of the ALP Opposition, 1951-1960.

the airline from utilising its self-insurance funds, while in 1973 TAA was given the power to broaden its activities into areas incidental to its airline operations. The most recent revisions, which were put before the Parliament in 1980 [DOT (1980a)], continue the effort to ensure that

> . . . the two airlines . . . as far as practicable, enjoy comparable cost structures.[66]

Hence one outcome of the TAP is that the TAA management perceives its goals to be similar to those of a private company. For instance, the then General Manager of TAA, Ryland (1967, p. 25) notes:

> . . . TAA in its day-to-day activity operates on the most vigorous of commercial enterprise lines.

Such similar management objectives, as well as the need to operate within the same regulatory environment, have important implications for the ownership question. Firstly, these factors tend to invalidate the metaphism hypothesis. In particular, the need for a government-owned firm to aid economic regulation becomes doubtful. Edwards (1969, p. 118) argues that state ownership is essential for influencing airline behaviour when sharp divisions exist between government objectives and normal commercial goals. Since Australian policy seeks to ensure that TAA behaves in a similar fashion to its private competitor, state ownership appears to offer little advantage for purposes of economic regulation. Even the ability to modify the behaviour of private companies through the threat of preferential treatment of the state firm is not a conclusive advantage of state ownership, since the existence of state ownership is clearly not a necessary condition for such preferential treatment (e.g. the position of ATI vis-a-vis other private Australian air operators).

Secondly, the possibility of the two operators performing to different standards of efficiency is reduced. In this regard it seems strange that Dedman was able to support the establishment of TAA on this very issue:

> It has been pointed out by many students of industrial organisation that there is very little difference between a large modern industrial company and a

---

66 Paltridge, *Hansard, S*, 28 September 1961, p. 721.

public enterprise itself.[67]

However, Davies (1971, 1977) argues that a state corporation is likely to be less efficient than a private firm due to an inability to transfer ownership.[68] Under private ownership some individuals have relatively large shareholdings (compared with the infinitesimally small share each person 'owns' of the government firm), and hence have increased incentive to maintain a check on the firm's performance. Individuals are also able to specialise by owning shares in firms for which they have a comparative advantage in knowledge and monitoring ability. Davies supports his argument by considering various labour productivity measures.

The impact of Davies' hypothesis might be expected to be weak in the airlines' case since, as noted above, Australia's stringent regulatory policies have been directed towards making the airlines as similar as possible. However, for this very reason, any observed differences can perhaps be considered more significant. Davies' empirical approach is also open to the criticism that the productivity differences which he observes may be due to factors other than ownership, e.g. scale effects, route structure and airline equipment.[69] However, some further evidence to support Davies' thesis comes from Mackay (1979) whose regression analysis takes into account many of the factors likely to influence airline costs and performance. His study indicates that, while both airlines have costs which are higher than might be predicted, TAA performs relatively less efficiently than the private company. For example, in 1975 TAA's costs were estimated to be 16.8 per cent higher than might be expected, while those of the Ansett airlines were 12.8 per cent higher. An overview of this ownership issue suggests that, while there appears to be some theoretical and empirical evidence that the state firm is less efficient than the private one, this difference is likely to be small compared with the inefficiences of both operators which are due to the current policies of economic regulation.

---

67 *Hansard, HR*, 25 July 1945, p. 4560.

68 Davies also recognises that a regulated private firm is likely to be less efficient than an unregulated one.

69 These issues are also discussed in Forsyth and Hocking (1980) and Davies (1980).

Problems induced by government ownership

The introduction of a state-owned company into an industry also presents some distinct problems. There are important difficulties in achieving equilibrium in an industry with both state and private operators. In the words of Menzies:

> . . . everybody in Australia knows, that if a government is in a business such as air navigation, and a private concern is in the same business in the same field, then the government can, subject to control or statute, destroy its competitor at its own sweet will.[70]

In the past the private airline operators had no reason to doubt this to be a realistic possibility under an ALP government. Evatt informed the Parliament:

> A future Labour government will frankly tell the people that the new parliament will not be bound by its predecessor.[71]

In addition, the private operators had already observed the preferential treatment given to TAA in the immediate post-war period. ATI (1980, p. 4A) lists some of the advantages which it claims TAA enjoyed; these include favourable loans, all Government business, no dividend requirements and favoured treatment regarding airport facilities and aircraft permits. DOT (1979a, p. 36) acknowledges that TAA then received major competitive advantages over its rivals. Even to the present, ATI's attitude towards airline regulation continues to be dominated by a perceived need to ensure protection from the actions of a hostile government [see ATI (1980)].

---

70 *Hansard*, *HR*, 30 October 1952, p. 3987.

71 *Hansard*, *HR*, 29 October 1952, p. 3883. Evatt's backbenchers were somewhat more blunt:

> As soon as the Labour Party regains power, and that will not be long delayed now, we shall run Australian National Airways Proprietary Limited out of the air and make Trans-Australia Airlines a Government monopoly. [D.J. Curtin, *Hansard*, *HR*, 22 May 1952, p. 789]

While the ALP's policy has since changed to one of:

> an evenhanded 2-airline agreement under which the two airlines will be operating on identical terms[72]

a fear remains that, for a given situation and no matter how unbiased are government actions, the private firm faces a greater risk of exit from the market than does the state firm. This fear is real[73] and seems reasonable. In these circumstances competition between state and private firms can never be truly 'fair'. This then is the 'Australian problem' for which Brogden (1968, p. 202) considers the TAP to be the valid and only long term answer.

Relevance of the ownership question

Many authors have recognised that the sale of TAA to private interests within the framework of the TAP would result in little change in the overall performance of the industry. Hence the ownership question is often seen as a 'red herring'.[74] However, Ansett (1979) notes:

> There would not be the same necessity for these provisions and facilities if the competing airlines were both privately owned and shared identical responsibilities to those of other public companies in this country.[75]

It seems then that, rather than being an irrelevant issue, the state ownership of TAA is a crucial political obstacle to open market competition in Australia's civil aviation industry.

---

72 Jones, *Hansard*, *HR*, 24 May 1973, p. 2365. Jones was Minister for Transport, 1972-1975.

73 '. . . should that airline [TAA] fail to achieve any profit its survival capability is unaffected.' [Ansett (1979)] Of course there may exist other private operators who are prepared to accept this additional risk.

74 *Australian Financial Review*, Editorial, 17 May 1979.

75 Note that this statement also implicitly dismisses the many other arguments raised in support of economic regulation of the airlines.

4
The Safety Issue

The concern for air safety

The issue of air safety has undoubtedly been an important consideration in both the debate on airline regulation and the civil aviation policy formulation of all Australian governments. The provision of safe air services has been a major objective of such policies. Drakeford considered:

> 'Safety must be the first consideration in air transport'[76]

while Nixon (1978, p. 13) urges that air accidents are to be avoided 'at all costs' as

> it is the Government's policy to maintain aviation safety at the highest possible level.[77]

A domination of aviation matters by safety considerations is supported by Hewitt (1979a, p. 4) who argues:

> no issue . . . is more important than operating safe airlines

and by AFCO (1978, p. 12) which holds the view that the primary role of government is to ensure safety.

Safety and economic regulation

Throughout the airline regulation debate many writers have attempted to link the safety performance of air transport firms with their financial positions and hence with economic regulation of the industry. In particular, a constant theme is

---

76 *Hansard*, *HR*, 6 November 1951, p. 1545.
77 *Hansard*, *HR*, 16 August 1978, p. 355.

the incompatibility of safety[78] with the profit motive and the competitive process.

Brogden (1968, p. 200) considers that 'the profit motive is not the vital factor' in maintaining high safety standards, while another contributor asserts:

> ... most aircraft accidents resulted from the anxiety of private enterprise to make more dollars regardless of the sanctity of human life.[79]

The relative lack of concern for profits by a state-owned firm was thus often thought to have favourable implications for the safety standards offered by that firm. This point is clearly made by Calwell:

> Ansett-ANA is run for profit. TAA is run for the benefit of the people. Both airlines conform to the requirements of the Department of Civil Aviation, but TAA does spend more on maintenance and is more careful. It goes further than ANA can afford to go as a private enterprise, profit-making organisation.[80]

Since the existence of profit-motivated private firms is likely to have a harmful impact on safety, then it allegedly follows that open market competition among such firms can only exacerbate the problem. Thus there is

> no question that 'rat race' competition can and does lower safety standards' [Brogden (1968, p. 200)],

and hence:

> the development of fierce competition between the two major airlines would lead to disaster in the future.[81]

---

78 The discussion is naively simplified by a tendency to categorise air services as either 'safe' or 'unsafe'. This approach is retained at this stage, but criticised below.

79 Haylen, *Hansard*, *HR*, 29 October 1952, p. 3892.

80 *Hansard*, *HR*, 30 September 1958, p. 1761.

81 Cole, *Hansard*, *S*, 1 October 1958, p. 778. Senator Cole was Leader of the DLP, 1955-1965.

The economic position of an airline operator has been claimed to affect air safety in several ways.[82] Firstly, firms may be undercapitalised or of weak financial strength due to insufficient profitability, with the result that they are unable to provide the level of facilities (e.g. workshops and crew training) which is necessary to maintain the accepted standards of safety. However, the discussion in Chapter 1 indicates stability of market structure in a competitive environment rather than the continual emergence of financially unsound firms. In addition, under the Air Navigation Regulations (ANRs) the inability of an operator to satisfy safety requirements is sufficient grounds for refusing to issue or for suspending an air service licence. Hence financially weak airlines, if considered a threat to air safety, could be readily eliminated through licensing procedures.

Secondly, a firm which finds itself in a situation of falling profits may be tempted to skimp on safety measures by, say, reducing maintenance expenditure or operating in marginal weather conditions. This argument can easily be overstated. Aircraft crew, despite the financial plight of their company, would still be likely to give heavy weight to their own personal safety (though differences in risk aversion between travellers and airline employees are theoretically possible). There may also be relatively little difference in the costs of flying an aircraft safely and flying it at all. Ramsden (1976, p. 37) mentions examples such as meetings of maintenance heads and voluntary incident reporting, and notes that these procedures serve both operating efficiency and safety so that it is difficult to distinguish between the two. Similarly, Edwards (1969, p. 49) recognises that the main safety effort on the part of air operators is 'inseparable from the day-to-day work'. Hence the absolute gain from skimping may be much smaller than is usually presumed.

In addition, the skimping argument ignores some counteracting competitive forces. Adequate safety is in the firm's own self-interest since, in this simple safe/unsafe scenario, safe operations appear to be a necessary condition for profitability. This implicitly assumes that there are insufficient consumers who would prefer the 'unsafe' service to the 'safe' service in order to make the former operations profitable. Poulton (1965, p. 15) thus claims:

---

[82] Hewitt (e.g. 1979a, 1979b) has recently been the most vocal proponent of the view that airline competition is incompatible with maintaining safety standards and discusses some of the mechanisms considered below.

> A safe industry is a prerequisite for a stable and profitable industry.[83]

The adverse effect of a lack of safety on an airline's profits may also derive from market responses more indirect than declining consumer patronage. For example, Ramsden (1976, p. 173) notes that the relative safety of airlines is reflected in insurance premiums for aircraft which may thus vary from 1-3 per cent of their replacement value. Hence in a competitive environment one might expect profitable firms to be a subset of those offering safe services. The profitable airlines will presumably be the ones best able to satisfy consumers' preferences for other service characteristics.

However, while this analysis implies that profitable firms will provide safe services and that unsafe firms will not exist under this simple regime of market competition, it does not necessarily follow that safety will be ensured by providing firms with guaranteed profits through economic regulation of the industry. Although some observers recognise that a viable system composed of financially sound operators can afford to devote resources to air safety,[84] there is no guarantee that greater attention to safety matters will actually be forthcoming in the absence of direct regulation of safety standards. AAA (1979, p. 7) implicitly agrees with this point when it notes:

> a financially stable airline is a potentially safe one . . . it can readily afford the highest standards of equipment, engineering and training.

Thus, under a system of only economic regulation, safe airlines are a subset of those made profitable by such regulation.

Ramsden (1976, p. 185) also rejects the argument that air safety is likely to be directly impaired by financial weakness through the methods discussed above. However, he suggests that financial weakness might have an indirect effect through 'subtle corporate incapacitation'. Under this hypothesis a deteriorating commercial performance may lead to a creeping management shambles which degrades the disci-

---

[83] Ramsden (1976, p. 167) agrees with this view:

> A safe airline is not necessarily financially sound, but an unsafe airline cannot be.

[84] E.g. Anderson (1973, p. 18) and Chippindall (1965). Sir D.G. Anderson was Director-General of Civil Aviation, 1956-1973.

pline and morale essential for safety and thus makes the operator vulnerable to accidents. However, Ramsden also notes a lack of statistical evidence to support the hypothesis and the observation of many counter examples where airlines have gone bankrupt without any sacrifice in safety. Furthermore, it would be unlikely that safety would be the only area affected by the alleged management shambles. One might also expect other quality aspects of the declining operator's air service to deteriorate, hence speeding its exit from the industry and minimising any consumer exposure to the supposed fall in safety standards. Finally, the administration of licensing procedures and other direct controls on operating standards can be expected to ensure that 'subtle corporate incapacitation', like the previous two methods, imposes little threat to consumer safety in a competitive airline market.

While many assessments of Australia's TAP regard our achievements of high safety standards as a benefit of the policy,[85] the existence of such a causal relationship between economic regulation and safety is tenuous.[86] Both Edwards (1969, p. 211) and Ramsden (1976, p. 185), two of the main proponents of a link between economic strength and safety, note that there is little empirical support for this relationship. Observation of the high safety levels achieved in the competitive Californian intrastate market and the explanation of the different safety records of the U.S. domestic airlines in terms of chance fluctuations [Barnett et al (1979)] provide further evidence on this issue. Similarly, Australia's impressive safety record can be more accurately traced to causes other than economic regulation; for instance, the ideal flying conditions due to favourable climate and geography[87] and the strict system of safety and operational controls [Brogden (1968, p. 200)]. Ramsden (1976, p. 20) agrees with this assessment. He attributes Australia's safety record to good weather, the general airmindedness of the

---

85 E.g. Nixon (1976, p. 39), Paltridge, *Hansard*, *HR*, 3 September 1957, p. 84 and *The Age*, Editorial, 28 December 1979.

86 This view is supported by Levine (1965, p. 1429), Miller (1975, p. 693) and Keplinger (1976, p. 191).

87 See Corbett (1965, p. 17), Wilson (1962, p. 46) and Hocking and Haddon-Cave (1951, p. 4). Interestingly, Butler (1971, p. 17) disputes this opinion. It might be expected that technological advances would tend to lessen the impact of natural conditions on aviation safety.

country and tough safety standards. He refers to Australia as a 'police state' in its air safety regulation and enforcement.

If the community decides that certain minimum standards of safety are necessary, then direct regulation of operating standards appears a more efficient means of pursuing that objective than a system of economic regulation.[88] However, the argument is sometimes put forward that direct safety regulation is more easily implemented when the industry is also subject to economic regulation [e.g. Chippindall (1965)]. The process through which this occurs is usually not explained. Even if the claim is true, it is not necessarily the case that the extra benefit is greater than the possible costs of economic regulation.

It follows from the above discussion that economic regulation of the airlines and considerations of air safety are two distinct issues. Hence the economic deregulation of airline markets would not necessarily prejudice air safety. Indeed, while many writers have called for the removal of market restrictions from the industry, they typically support the retention of direct safety controls and the maintenance of current standards. There has been little questioning of the need for these safety regulations.[89] DOT (1979a, p. 57) notes:

> There was no suggestion in any submissions that safety regulatory control in relation to trunk services should be varied and the Committee sees no need to make any changes in this area.

But why is government intervention necessary to ensure adequate safety standards? Why does the market fail to provide the desired standards? A brief discussion might clarify some of the essential issues in this highly emotive area.

---

88 A system of direct regulation of safety standards is itself not costless; hence these costs must be weighed against the benefits. The economic nature of air safety is discussed further below.

89 A rare exception is BTE (1978, p. 59):

> Certainly, there is scope for conjecture as to whether existing safety standards may be too rigid in an economic sense (and it could be that they may be counter-productive in some instances).

Towards an economic analysis of air safety

There is no reason why air safety cannot be considered an economic good: while consumers positively value higher levels of safety, it also costs extra resources to produce. Hence, as with all economic goods, there will be an optimal level of safety, a level beyond which consumers judge the extra safety to be gained not to be worth the cost in terms of the value of alternatives foregone. If this were not so and if the ultimate objective were safety 'at all costs' or 'at the highest possible level', the obvious policy solution would be to ban civil aviation. The apparent absurdity of this policy prescription merely reflects the inappropriateness of attempting to pursue such narrow goals.

The economic nature of air safety is illustrated by many examples in which safety is compromised by other economic considerations. The system of permissible unserviceabilities enables operators to correct some defects at a more convenient time and place. Ramsden (1976, p. 75) notes that passenger survivability of airline crashes could be expected to increase with the introduction of upper-body seat belt restraints or rearward-facing seats. However, consumer acceptability is a major influence preventing the introduction of such safety measures (especially for the latter innovation which appears to have minimal cost implications). In Australia most commuter operations are conducted under an exemption from the need to satisfy the usual technical standards relating to scheduled air services. The reduced safety standards implicit in these exemptions effectively lowers the costs of operation and increases the economic viability of such services. Another example is the F27 service of EWA to Norfolk Island from Sydney. Since February 1977 this service has been operated under a concession from the accepted safety standard for overwater operation by a twin-engined aircraft, namely, that in the event of engine failure, it must not be more than 90 minutes normal cruising from a suitable aerodrome. The EWA service exceeds this maximum by about 15 minutes.

Recognition of the economic nature of safety illustrates the misleading approach of categorising air services as 'safe' or 'unsafe'. There exists a broad spectrum of safety standards, ranging from high to low in terms of the probability of suffering injury or damage. In addition, different consumers with different budgets, tastes and attitudes to risk would be expected to prefer to purchase different amounts of air safety.[90] Hence without safety regulation one would expect to observe different operators offering different safety standards to satisfy the different groups of consumers. It follows that a system of direct safety regulation which truncates the distribution of available safety choices by setting minimum standards does not necessarily have any a priori merit.

Nixon (1978, p. 13) leaves little doubt that considerations of the public interest are not the main concern when determining safety requirements:

> . . . the general aviation industry must realise that while aircraft accidents continue to occur, there will be public demand for more and more regulation, irrespective of the merit of such regulations.

Further doubts on the desirability of safety control come from the suspicion that such legislation is a possible means of extracting economic rent from the market. Safety standards are not created in a vacuum but rather derive from the interaction between the bureaucracy, airlines, airline employees and consumers, each group with a different degree of influence. Higher standards may in fact be a convenient means of increasing job opportunies and rewards for airline employees or of giving bureaucrats greater power and prestige. An example is the relatively high standards relating to aircraft evacuation which sometimes require Australian airlines to carry more cabin crew than overseas operators [see ATI (1980, p. 13)]. Many Australian standards are recognised as amongst the strictest in the world. Similarly, Hewitt (1979a, p. 4) warns of the threat to air safety which would occur if airline management were subjected to the same financial pressures as the 'harried' garment industry. However, it should be noted that a central feature of his strategy to ensure air safety involves the lessening of market pressures and disciplines facing airline executives.

In addition to these doubts on the standards which have been adopted in the past, one should also be aware of the various economic responses to the introduction of compulsory minimum standards. These responses may limit the capability of policy to achieve public interest goals. For

---

90 This should come as no surprise; not everyone chooses to buy the 'dynamic safety' of a Volvo, or to use a car instead of a motor bike, or not to play body contact sports, etc.

example, stricter air safety standards may, by increasing the cost of air travel, induce consumers to switch to other modes of transport (e.g. private cars) which may have higher safety risks. Thus the value of the reduced risk to continuing air travellers must be offset by the exposure of previous air passengers to the lower safety of the chosen alternative mode of transport. A priori, it is not clear that the community will necessarily enjoy net gains from the increased air safety standards.

Any economic justification for government involvement in the determination of safety standards lies in the existence of externalities or market failure. The most likely source lies in the provision of information on the safety standards of individual operators. Other possible sources of market failure include safety spillover effects (a poor safety performance of one operator may generate a lack of confidence in the air system as a whole [e.g. Edwards (1969, p. 10)]) and the danger from air accidents to residents (though this danger could be reflected in housing prices and hence be implicitly accepted by consumer decisions on residential location). Both of these potential sources of market failure also seem closely related to the provision of consumer information. The above discussion implicitly assumes that the consumer possesses the necessary information regarding safety to make his rational decision. Thus, when the debate turns to the question of skimping on safety, the real worry is not the skimping per se but rather the likelihood that the consumer will be ignorant of such actions and hence might be purchasing a product which is not what he thought it to be. However, the most appropriate role for government in these circumstances appears to be to directly correct the informational deficiencies, by perhaps instituting a system of voluntary certification of services and allowing fair advertising of these product characteristics as an alternative approach to the setting of compulsory minimum standards or economic market control.

Relevant to the question of advertising air services is the observation that safety has traditionally been 'that most taboo of airline subjects' [Hewitt (1979a, p.4)]. This situation is unsatisfactory when the issue is frequently used by advocates of economic regulation of airline markets. One hypothesis of the reluctance to discuss the safety performances of different airlines might be that, after allowing for various special factors (e.g. climate, terrain, average stage length and aircraft type), differences in air accident statistics may be chiefly determined by chance. A reticence to discuss the special factors influencing an airline's safety performance can be contrasted with the numerous reasons which are advanced to explain differences in economic performance [see Chapter 6].

Summary

This chapter's discussion on safety and economic regulation of airlines can be neatly summarised by the following:

> Arguments about safety standards and safety records are not relevant in this regard. Safety standards can and do coincide with lower fares and more competition between airlines. After all, the maintenance of safety standards is one of the most effective ways of competing. The safety issue is simply a bogy which is trotted out to frighten the naive whenever the airline monopolies are under threat. [91]

---

91 *Australian Financial Review*, Editorial, 11 January 1979.

5
Some Social Objectives

Defence

It is not surprising that the air transport industry is closely associated with national defence considerations. The first major application of aircraft was for military purposes and subsequent conflicts provided further stimulus to civil aviation through technological advances, the availability of trained personnel and by boosting the demand for air services. From its earliest years, not only did many operators and bureaucrats have military flying backgrounds [see Goodrich (1960, Ch. 1)], but also, as Butler (1971, p. 31) reports, no person could be employed as a pilot, ground engineer, navigator or wireless operator unless he agreed to serve in the RAAF reserve. Chippindall (1965) warns not to overlook the importance of the airline industry for a future war emergency. Thus, when discussing aviation policy:

> Defence considerations must also be taken into account here, bringing with them the need for continuing large Government role for aviation in Australia.[92]

While the defence role of military aircraft is obvious, the corresponding role for civil air operators is less so. Although the planes of these operators can be and have been pressed into military services, it is generally agreed that the chief value of airline operators in a defence emergency lies in the effective performance of civil aviation tasks. Goodrich (1960, p. 71) recognises this point:

> It may be agreed that it is important for national defence for Australia to have a large and efficient domestic and international air transport system . . . Civil aviation undoubtedly has an emergency trans-

92 *The Age*, 21 July 1966, p. 13.

> portation role for troop movement, civil evacuation and the like, but it should not be regarded as a cut-rate method of obtaining and maintaining transport aircraft and personnel for the Air Force to take over when hostilities commence. Civil air transport has its own vital contribution to make by providing efficient speedy transport for a nation at war.

Chippindall (1965) also stresses the extra demands placed on the air transport sector to perform civilian air services during an emergency situation.

Civil aviation, like many other industries, undoubtedly becomes more important in times of war since the demand for air services increases greatly. But, increases in demand also occur during peacetime, though perhaps less dramatically. It seems then, that the most appropriate policy, in order to satisfy the actual and potential demands placed on the industry, is to provide the necessary infrastructure and to encourage the growth of the most efficient operators. One would expect an economically efficient industry to be more capable of showing the flexibility and innovation necessary to cope with a rapidly changing wartime economy.

One of the stated aims of the Civil Aviation Agreement Act 1952 is 'to assist the defence of the Commonwealth'. However, it is unlikely that the TAP is the most effective means of achieving civil aviation defence goals. McMahon argues that an adequate defence potential can be maintained 'by the continued existence of the 2 main airline operators'.[93] Others stress the dangers of competition to Australia's defence:

> Air transport is too vital to Australia's economic progress and defence security for it to be open to the risk of having its scope and efficiency impaired by a wasteful rat-race between uncontrolled private competitors.[94]

However, the discussion in Chapter 1 indicates that competition will not destroy the airline industry's ability to satisfy consumer demand and hence its defence capability. McManus emphasises the duplication of facilities under the

---

93 *Hansard*, *HR*, 30 October 1952, p. 3979. McMahon was then Minister for the Navy and Minister for Air and subsequently Prime Minister, 1971-72.

94 *The Age*, Editorial, 23 July 1966.

TAP and the availability of additional trained staff.[95] Given the discussion on economies of scale in Chapter 2 this effect is likely to be insignificant. A further defence advantage often attributed to government involvement in air transport is the co-ordination of civil and military aviation policies. This was one of the reasons offered by Drakeford to support the ALP's attempted nationalisation of the airline industry.[96] Such co-ordination is supported by *Australian Transport* (1977, p. 17):

> Although the costs of the civil transport structure should be obtained from outside the defence vote, there is a need for all transport developments to be examined by defence with a view to providing extra fitments, . . . etc. in order to make the unit of transport, or the infrastructure, more readily convertible to defence needs. Such additions should be funded through the Defence Department.

In reality there is little such co-ordination under the TAP; *The Canberra Times*, 1 August 1979 reports the Secretary of Defence and the Chief of Defence Force Staff to claim that the domestic airlines do not consult with defence authorities when purchasing new equipment. Finally, to the extent that the current operators are not efficient [see Chapter 6], then the TAP seems to be detrimental to Australia's defence.

National development

Many observers stress the role which air transport plays in the national development of the country. Chippindall claims:

> Aviation . . . is in my view of such enormous importance to the development of our country that the Government must always have a direct role in the control and development of the industry.[97]

---

95 *Hansard*, *S*, 20 November 1957, p. 1397. Cole, *Hansard*, S, 1 October 1958, p. 778 takes this argument to its limit by claiming that 'the more airlines we have the safer we may feel'!

96 *Hansard*, *HR*, 18 July 1945, p. 4184.

97 *The Age*, 21 July 1966, p. 13.

TAA argues that the ability of domestic airlines to direct traffic to certain areas often provides a stable influence which attracts investment. Hence it claims:

> The domestic airlines have made and are continuing to make a substantial contribution to regional development and decentralisation.[98]

However, it is unlikely that the direction of causation is so simple; i.e. do airline services enable the development of new regions or vice versa? Edwards (1969, p. 183) admits that there is almost no empirical evidence to support the former proposition. On this point one might also note the expansion, especially in Western Australia, of demand for and thence supply of air services which has been associated with the recent development of resource projects.

Holt (Prime Minister, 1966-1967) attributes the benefits more directly to the TAP, through which

> We have fostered the development of outback services, and in this way, assisted national development in remote areas.[99]

However, once more it appears that the attainment of any chosen national development objectives is quite independent of the existence of the TAP. Such goals, if desired, can readily be subsidised through the cost recovery program[100] (as the airlines and others demand it should be)[101] and

---

[98] *Hansard, SCT*, 8 February 1977, p. 687.

[99] *Hansard, HR*, 9 April 1957, p. 628.

[100] The Commonwealth Government is largely responsible for the provision of airports and airways facilities. Part of the costs of these services are recovered from the aviation industry through its system of Air Navigation Charges (ANCs).

[101] E.g. Ansett (1972, p. 39) and Australian Federation of Travel Agents (AFTA), *Hansard, SCT*, 29 June 1977, p. 2241. However, DOT (1979e, p. 162) is unconvinced of the desirability of subsidising civil aviation on either defence or national development grounds. It argues that many industries and individuals play important roles in these fields and that there do not appear to be any cogent reasons why aviation should be given special treatment.

through direct subsidy payments for specific air routes, without the need for restricted market entry.[102] A preferable approach to national development goals is to adopt a policy producing the most efficient operators and encouraging the growth of developmental routes by direct subsidies through, say, a system of competitive tendering. A tendering system for allocating Government air subsidies was used during the 1920s and 1930s, although the choice of operator was not always based on the lowest tender [Goodrich (1960, Part I)]. DOT (1979a, p. 192) recommends granting air transport assistance to communities through the provision of explicit subsidies in the form of contracts to supply specified services. Such contracts would be awarded for a relatively short term with the operator selected by a process of competitive bidding.

Fuel conservation

Even the popular issue of fuel conservation has recently entered the air debate. It has been argued that with increasing costs of fuel a monopoly airline would be more efficient than the present Two-Airline system,[103] while Hewitt (1979b, p. 11) predicts that deregulation will not be tolerated in an energy-conscious world.

Elementary economic principles again suggest that it is unwise to pursue such a narrow goal as fuel conservation. Empty seats are not necessarily a waste but reflect a higher quality of service through more frequent flights, and the consumer may be willing to pay for this level of quality. Furthermore, the discussion in Chapter 1 suggests that competitive market forces will act to satisfy consumer preferences and to successfully adjust output and product characteristics to any changing market parameters, e.g. fuel prices. If fuel conservation is considered a problem, the first policy requirement is an economically sensible strategy for pricing fuel, not detailed regulation of industries using fuel as an input.

---

102 The practice of encouraging a greater geographic spread of services through cross-subsidisation methods is criticised in Chapter 1.

103 *The Canberra Times*, 11 October 1979. Similarly, Thayer (1977, p. 684) considers fuel conservation to be the new rationale for 'some kind of global public utility system'.

Free enterprise values

Mention should also be made of another set of social objectives: those associated with free market or private enterprise values. The value of individual decision making and freedom of choice and action as well as

> the values of not having businessmen dependent on grants of privilege . . . with its attendant risks of mutual corruption of both the political and economic processes [Kahn (1971, p. 246)]

are highly regarded by some. For example, AFTA remarks:

> It is believed that this policy has created a cartel and is contrary to the public interest as well as being out of place in a free enterprise society.[104]

To the extent that such values exist, economic regulation of the airline industry imposes additional costs on the community.[105]

A special industry

A notable feature of the Australian airline regulation debate is the regard by many of aviation as such a special industry that government involvement is mandatory. Hewitt (1979a, p. 5) considers the proposal that the airline industry is no different, and hence should be treated no differently, to other industries to be 'a fallacious and dangerous point of view'. Drakeford argues:

> . . . aviation is intensely vital today as a national instrument, and that inevitably it is a part of the very core of the economic, social, diplomatic and defence policy of the nation.[106]

---

104 *Hansard*, *SCT*, 29 June 1977, p. 2237.

105 This must be weighed against the benefits to those people who genuinely value regulatory processes as an allocative mechanism.

106 *Hansard*, *HR*, 18 July 1945, p. 4179.

In a similar vein it is argued that civil aviation has

> a value away and beyond what it costs to operate in dollars and cents[107]

and that

> the so-called vested interest in air transport, far from being confined to the travelling public, belongs to the entire community.[108]

While airlines undoubtedly perform a valuable task, so, by definition, does any firm which sells a product in the market place. Unless there is a marginal benefit or cost which is not reflected in the price of the service (as the previous two quotations assert), there seems little need for government involvement. A common theme throughout this chapter is that the social objectives most frequently raised in the context of civil aviation can be addressed independently of restrictive economic regulation of the industry. Furthermore, to the extent that the current regulatory policy leads to economic inefficiency in the airline industry, the pursuit of such goals is likely to be hindered. Finally, one suspects that the significance of these external effects can often be exaggerated and that hence the operation of airline services is largely a private good in nature:

> The argument has been advanced that aviation provides a social service and therefore should be subsidised. This view is not acceptable. There are better ways of providing social services than through heavily subsidising aviation . . . Air transport is clearly an economic service and, as such, there should be the requirement that the user or beneficiary meet the cost of the services provided. [Jones (1975, p. 18)]

---

107 Captain R. Holt (President of the Australian Federation of Air Pilots), *Aircraft*, May 1977, p. 10.

108 Drakeford, *Hansard, HR*, 18 July 1945, p. 4181.

6

The Two-Airline Policy: Past and Future

Any judgement of the success or otherwise of the current regulatory framework, the need for policy reform and the choice among the various policy prescriptions must be based on a sound assessment of the industry's economic performance. While there may be disagreement about this performance, there can be little doubt that it is largely due to the system of regulation adopted for the industry [see Appendix 1]. This chapter surveys the airline industry's performance under the TAP and examines some of the proposed options for future aviation policy.

I. THE NEED FOR POLICY REFORM

A favourable viewpoint

Many observers consider the TAP to be successful, producing a number of desirable results for the community. This is not only the opinion of the two major operators;[109] several politicians,[110] bureaucrats[111] and others[112] offer the same view. Perhaps Cotton presents the most glowing account of Australian airline regulation:

> I think it has been acknowledged by all people that the 2-airline policy has played a major role in Australia's development and has maintained stability . . . the airlines have achieved a high standard of service . . . The 2-airline system is the envy of many

---

109 E.g. Ansett (1965, 1972) and TAA, *Hansard, SCT*, 8 February 1977, p. 706.

110 E.g. Swartz (1967, 1969). Swartz was Minister for Civil Aviation, 1966-1969.

111 E.g. DOT, *Hansard, SCT*, 21 March 1977, p. 1125.

112 E.g. Brogden (1968, p. 207), Wettenhall (1966, p. 72), Hocking (1972, p. 55) and the Editorials of *Aircraft*, August 1961 and August 1965.

countries . . . I think it is one of the most admirable regulatory air transport systems in the world . . . I think the system has demonstrated that it stands alone in its ability to provide the Australian public with what is undoubtedly one of the safest and most efficient air services in the world, and whose fares are as low as any in the world when one considers all the factors involved.[113]

The discussion contained in previous chapters indicates that many of the desirable features currently available in the industry (e.g. safety, reliability, stability and modern equipment) would not necessarily disappear in the absence of the TAP. A further worrying aspect of the favourable viewpoint is the frequent use of excessively narrow assessment criteria. For instance, Chippindall (1965) draws attention to the growth in passenger traffic over time as evidence of the success of the TAP and of public satisfaction with the policy. A more relevant consideration is the growth likely to have occurred over that same time period if a different regulatory system had been adopted. Similarly, Watkins (1963) judges that the TAP is effective as the major operators share such a large proportion of the domestic market. A common shortcoming is to assess the TAP solely with respect to its impact on the airline operators (e.g. their stability and viability) with little consideration given to its effect on consumers.

The recent economic critique

While the TAP has attracted criticism throughout its history, the late 1970s saw a renewal and consolidation of dissatisfaction with the policy, particularly in the form of more rigorous economic analyses of the system.[114] This research was undoubtedly stimulated by advances, occuring mainly in the U.S., in the theoretical and empirical knowledge of economic regulation in general and of airline economics in particular.

It is generally acknowledged that the Australian domestic airline industry has been characterised by a lack of substantial competition. When the TAP was first introduced

---

113 *Hansard*, *S*, 26 October 1972, p. 2301.

114 E.g. Forsyth and Hocking (1978), BTE (1978), Kirby (1979), Forsyth (1979) and DOT (1979b), particularly the appendixes of Mackay, Hocking and Gannon.

Evatt noted:

> This proposed rationalisation scheme is the very antithesis of active and competitive air services . . . In fact, for all practical purposes, the agreement will put an end to real competition . . .[115]

The regulatory system gives the airlines not only the opportunity but also the encouragement to reach mutually acceptable agreements, i.e. to collude. When outbreaks of competitive behaviour have occurred, they have largely been confined to peripheral areas such as food and drinks, staff appearance and uniforms and exaggerated advertising claims. The recent self-dubbing of AAA as 'the competitive one' might fall into this last category, although some observers could consider the slogan to be more a case of misleading advertising! Some of the most active competitive battles have been fought in the courts and in the Rationalisation procedures of the TAP.

For an example of this last point consider the saga of TAA's DC9 service on the Perth/ Port Hedland /Darwin route.[116] TAA was granted a licence to operate this service in February 1974. ATI undoubtedly expected such a service to have an adverse effect on its regional airline, MacRobertson Miller Airline Services, which already operated over the same route with F28 aircraft. Unable to reach an acceptable agreement with TAA on this matter, ATI referred the dispute to the Rationalisation Committee in March 1974. Since the two airlines were still unable to agree, the Co-ordinator decided in April that TAA should be given access to a weekly service on the route. Dissatisfied with this decision ATI appealed to the Arbitrator who in June reaffirmed the Co-ordinator's decision. Having for the time being exhausted the Rationalisation process, ATI resorted to the legal system, arguing before the High Court in June 1974 that it was not valid under the Australian Constitution for TAA to offer intrastate services on the Perth/Port Hedland segment of the route. Apparently growing tired of these delays TAA indicated in August its intention to introduce the service while awaiting the High Court ruling; injunctions issued by the Victorian Supreme Court prevented this action. In December 1976 the High Court ruled in favour of the proposed TAA service. Following further procedural

---

115 *Hansard, HR*, 29 October 1952, p. 3877.

116 Outlined in DOT, *Annual Report*, 1977/78, p. 173.

manoeuvres the Arbitrator resolved some outstanding matters regarding timetables and fares in October 1977. TAA was able to commence operations 'at long last'[117] in November 1977. *Aircraft* (November 1977, p. 12) estimates that the delay to TAA was worth $1m annually to ATI.

A counter-argument has been presented which claims that the industry has, in fact, enjoyed vigorous competition which has resulted in a satisfactory economic performance by the airlines and which explains the similar prices and services observed in the industry. Airline operators and regulators both place heavy emphasis on the extra revenue which results from a favourable one per cent market swing and claim that this provides a competitive spur.[118] However, since the regulatory framework effectively limits the scope for gaining increased market shares and is directed towards ensuring that both operators remain viable and perform comparably, this spur is not likely to be particularly sharp.

The capacity determination procedures of the TAP, which control the total aircraft capacity available and which provide for each operator to offer half of this total, have a strong influence on the market shares of the two operators. However, it is not clear exactly to what extent these controls limit the attainable market share of competitive route traffic. Freeland (1977, p. 7) claims the range is 48-52 per cent of the market; AFCO (1978) suggests the airlines can compete for only 10 per cent of the market. If the load factor were 65 per cent on all flights on all routes and total demand remained unchanged, the market share which is physically possible ranges from 23 to 77 per cent. However, the economically possible market share will reflect the decreased attractiveness of an airline when it has higher load factors, and hence will generally be substantially less (especially when all the other important service characteristics are identical). Six-monthly data for the five years to the end of 1978 reveal that TAA's share of competitive route passenger traffic has varied within the range of 48.8-51.1 per cent.[119]

In a regulatory environment which is directed towards ensuring the stability and viability of the existing operators and which effectively limits the ability of each to make significant market gains at the expense of its competitor, the

---

117 TAA, *Annual Report*, 1977/78, p. 7.

118 See Chippindall (1965), Ansett (1965) and Freeland, *Hansard, SCT*, 17 August 1978, p. 5592.

119 Data from DOT, *Annual Report*, various years.

airlines may prefer to accept the 'quiet life'. Within the current regulatory framework such a reaction need not even be confined to only tacit collusion. A lack of real competition in the airline industry is alleged to have resulted in both allocative and technical inefficiencies.

Allocative efficiency

The TAP provides little incentive or opportunity to innovate. Rationalisation procedures form an obstacle to innovative attempts to expand the market size and to increase an operator's market share. Kahn (1979, p. 8) likens this sort of regulatory process to 'a patent system in reverse': unlike the patent system which rewards an innovator for the public disclosure of his new idea, Rationalisation either forces the innovator to share the benefits with the other airline or enables its competitor to block the opportunity to test it in the market. By effectively reducing the expected return from the development of innovative air services, the current regulatory system is likely to lessen the amount of resources that the airlines devote to such activities. Similarly, an average cost pricing rule discourages the airlines from offering services which involve a greater degree of risk than that implicitly contained in the allowed rate of return. In addition, rigid application of a 65 per cent load factor when determining capacity means that any innovative attempts such as discounting to attract more travellers are self-defeating; average load factors are held constant and revenue is diluted [see Hocking (1979b, p. 95)]. However, this last problem has recently eased to some extent: standby passengers are not counted for capacity determination purposes thus enabling average load factors to rise.

Allocative inefficiency is reflected in the absence of discounting and the small variety of price/quality choices available to the public. DOT (1979a) concludes that the overall level of fare discounting in Australia exceeded only that of the Third World domestic airlines, that this indicated a failure to adequately satisfy consumer preferences, and that significant efficiency gains would result from a wider range of combinations of prices and qualities of service. Forsyth (1979, p. 68) regards the lack of price/quality variety as 'arguably the single most important objection to the Two Airlines Policy'.

Technical efficiency

Possibly even more damning to the TAP is the claim that the airlines are technically inefficient. This hypothesis is ultimately based on the elementary proposition that an industry is likely to be less productive if it is organised as a closed market rather than as an open market. One imagines that the longer the market closure and the less the inter-firm rivalry within the closed group, the greater will be the likelihood of technical inefficiency. This technical inefficiency is effected through the system of cost-plus pricing of airline services. The empirical evidence offered to support this hypothesis consists of fare comparisons among world airlines, partial productivity analyses [e.g. Forsyth and Hocking (1978)] and regression cost models [Mackay (1979)].

This evidence is not accepted unanimously. For instance, figures are often quoted to show that Australian air fares have fallen in real terms, that changes in air fares compare favourably with changes in other consumer prices, and that air travel is now less expensive in terms of wage earnings [e.g. AAA (1979a) and TAA (1978)]. Such observations, while perhaps interesting, are of little value for assessing airline performance. They are more appropriate as indicators of movements in relative prices or living standards. The crucial test is to compare actual fares and costs today with their potential today under a different regulatory framework.

However, direct fare comparisons and partial productivity analyses are open to the valid criticism that they fail to consider the different operating environments of the various airlines which are being compared (e.g. input prices, scale effects, route density and average stage length).[120] These special factors need to be considered carefully. For instance, the domestic airline operators frequently cite the relatively high fuel tax they pay compared to overseas operators. But, since these payments are attributable revenue for cost recovery purposes, it is the Government's cost recovery program rather than the fuel tax itself which influences airline costs. If the airlines did not pay fuel tax they would pay higher Air Navigation Charges. Thus, within the context of an unchanged cost recovery objective (the most sensible ceteris paribus assumption), the fuel tax issue is largely irrelevant to airline costs. Similarly, the impact on costs of higher input prices and technical flying standards

---

120 See AAA (1979a, 1979b), ATI (1980) and TAA (1978).

should not necessarily be accepted at face value. These are determined within the regulatory environment and so may be partly influenced by that regulation. For example, unions may be able to secure higher wages in a regulatory framework which involves cost-plus pricing.

This issue of 'regulation exploitation' is discussed by Forsyth and Hocking (1978, p. 13) and mentioned in Chapter 4 in the context of safety standards. However, ATI (1980, p. 29) denies the possibility of such exploitation, arguing that the profit motive mitigates against it and that the demands of Australian industrial unions would not be particularly affected by the industry's viability or its price setting methods. It has already been noted that in the current regulatory environment the role of the profit motive is significantly muted and hence offers less protection from technical inefficiency than it does in a competitive market setting. Kirby (1979, p. 108) argues that, because of pressures to avoid the visibility of supernormal profits, raising the cost structure is one of the few means within the TAP framework by which the major operators are able to capture some of the monopoly rents which are potentially available from the entry restrictions. In addition, it seems implausible to suggest that the tenacity with which union demands are pursued is not affected by the reduced incentives of management to resist such claims.

To further support its case ATI (1980, p. 49) cites the refusal of the operators to allow three-man crews of DC9 aircraft in the mid 1960s.[121] However, since most other world airlines, the U.S. authorities and DCA firmly supported two-man crews, the major Australian operators could hardly have been in a stronger bargaining position, nor would their submission on this issue have been more obviously out of step with overseas practice and costs. A counter example to the DC9 case is provided by *The Canberra Times*, Editorial, 24 March 1980 which notes that, following industrial unrest due to amendments to Victorian workers' compensation legislation which limited the liability of employers, Ansett was one of the first companies to announce an agreement with the Victorian Trades Hall Council which provided that its workers would receive the same benefits as provided under the old legislation. The Editorial remarks:

> It would be harder to imagine a more dramatic demonstration of the costs that protected and

---

[121] See DOT, *Annual Report*, 1967/68 for further details.

uncompetitive industries can impose on the rest of the economy.

The regression model approach of Mackay (1979) attempts to allow for the effects of various exogenous influences on airline costs. He estimates that appropriate changes in the organisation of domestic air services might reduce average costs of production by as much as 35 per cent. Using this estimate Kirby (1979, p. 114) determines an upper bound for the welfare loss imposed on the community through technical inefficiency to be of the order of $250m in 1976/77, or approximately 50 per cent of gross industry revenue.

This regression technique must be acknowledged to be imperfect,[122] just as any attempted price justification procedure within the regulatory framework must also be. This stems from the difficult nature of the problem: the need to compare the actual performance of the industry with its potential. The most accurate test of the airlines' current performance is likely to be achieved only with open market conditions where other operators have the chance to demonstrate superior efficiency. Edwards (1969, p. 120) admits that regulatory devices such as efficiency audits and direct interference can never be as effective a spur to efficiency as the prospect of losing business to competitors.

Summary

It is clear that recent theoretical and empirical economic research has thrown an unfavourable light on Australia's TAP. The prevailing judgement of independent observers can be represented by the following:

> There is little doubt the two-airline policy as it operates now is strongly inimical to competition, cheaper airfares, and cost efficiency in both the major airlines.[123]

---

122 DOT (1979a) is critical of the regression technique and is reluctant to accept its findings. See Kirby (1980b) for a rejection of its criticisms and further discussion.

123 *Australian Financial Review*, Editorial, 11 January 1979.

II. SOME POLICY OPTIONS

No change

Several arguments can be advanced to suggest that there is no need to alter Australia's air transport policies. Firstly, it might be claimed that the TAP has worked well, producing favourable results and revealing few shortcomings. It is in this vein that Chippindall (1965) concludes:

> . . . any change at this juncture, even if it were practicable, would not be in the interests of the Australian people.

The discussion of the previous section indicates that this rationale for maintaining the current system is suspect.

A related argument stresses the allegedly 'theoretical' nature of much of the criticism of the TAP. Thus AAA (1979, p. 2) complains of 'nebulous suggestions of lower fares, improved services and superior timetables' and the lack of 'concrete methods by which these aims could be achieved'. Similarly, McKenzie (then General Manager of TAA) is unimpressed with 'classroom research' which is critical of the domestic airline system.[124] If such objections are to have any substance one needs to demonstrate the failure of economic theory in general [e.g. the attempt of Thayer (1977)] or its incorrect application to the airline industry. Yet the evidence indicates that economic analysis of civil aviation has successfully offered many insights into the airline industry and the effects of its regulation. Furthermore, the arguments for changing the current patterns of economic regulation are not confined to the world of theory. In particular, the U.S. deregulation experience provides important new empirical evidence to support the case for an increased reliance on competitive market forces. For example, *Australian Financial Review*, 20 February 1980 reports Meyer (president of Trans World Airlines) to note that U.S. deregulation has done for the airlines exactly what some economists predicted: forced the industry to become more innovative and efficient through competition.

---

[124] *The Age*, 16 January 1979.

An 'enlightened' approach

Several observers of Australia's domestic airline industry have proposed recommendations for policy reform which are based on perceived inefficiences in the TAP and seek to overcome these deficiencies. One broad class of such reform is the notion of an 'enlightened' TAP. This approach, while recognising current shortcomings, assumes that better administration of the industry within the basic TAP framework (most importantly, continued restricted entry) can achieve increased economic efficiency.[125] The modified policy typically seeks to encourage more competition within the TAP (e.g. by eliminating the more blatant collusive features such as the Rationalisation Committee), to introduce more effective regulatory scrutiny of the airlines (e.g. public fare inquiries and tighter cost examination), and to offer increased attention to consumer interests (e.g. by establishing an airline users committee or conducting passenger surveys).

Unfortunately one cannot be very optimistic about the likely success of this approach in achieving significant gains in economic efficiency. Since an 'enlightened' TAP does not appear to significantly change either the incentives facing the main groups of participants in the regulatory process or their relative powers of influence, it is doubtful that this policy response will yield the substantial efficiency gains which are potentially available.

The moves to encourage competition, while increasing the costs of collusion and enhancing the ability of the airlines to compete, appear to have only a marginal impact on their incentives to do so. Thus, with little threat of outside entry, a probable scenario is lengthy periods of tacit collusion interrupted by brief flurries of more visible rivalry, e.g. food and drink wars. Even when such competitive activities

---

125 DOT (1979a) is the most notable example of this approach. For a critical examination of this document see Kirby (1980b). An 'enlightened' TAP also appears to be the basis of ALP domestic civil aviation policy [see Morris (1980a)] and of the Government's recently announced policy initiatives [see DOT (1980a)].

125 DOT (1979a) is the most notable example of this approach. For a critical examination of this document see Kirby (1980b). An 'enlightened' TAP also appears to be the basis of ALP domestic civil aviation policy [see Morris (1980a)] and of the Government's recently announced policy initiatives [see DOT (1980a)].

occur, consumers will still be denied access to potentially lower cost firms (as will be the case if the airlines are currently operating from a technically inefficient base), as well as being limited to the innovative ideas of only the existing operators. The fundamental problem is to stimulate innovation and to ensure real competition; tinkering with the present system does not appear to be the solution. As Brogden (1968, p. 200) notes:

> A modicum of competition might be achieved by some adjustment of the controls to permit some greater degree of decision by the individual airline within the two-airline policy.

The idea that the policy modifications will enable regulators to be more effective presupposes that something is presently hindering their attainment of public interest goals. Tighter controls of costs and fares, while easy to recommend, are inherently difficult to achieve. For example, Forysth and Hocking (1978, p. 31) outline the informational requirements necessary for an idealised system of adjusting air fares. These requirements are so demanding that the prospects of achieving such an efficient regulatory pricing system are negligible. However, calls for public scrutiny of the regulatory process implicitly suggest that the major difficulty is a lack of incentive on the part of the regulators rather than any significant lack of ability. Morris (1980a, p. 5) complains that regulatory decisions are determined by a 'private club of industry operators and departmental bureaucrats'. He asserts:

> . . . the only effective way to ensure the public interest is protected . . . is by way of a public inquiry.[126]

Finally, most consumers are likely to gain little from surveys and institutional systems of consumer representation. What matters most is not what consumers say but rather how they vote with their purchases of airline tickets. A major danger with an institutional approach is that it may merely create another effective minority lobby group which is capable of further exploiting the general community through any revised patterns of regulation. In this regard consider the large number of special interest groups who have

---

126 Morris, *Hansard*, *HR*, 28 February 1978, p. 252.

recently been pleading their cases for air transport subsidies to the Independent Public Inquiry into Domestic Air Fares.

An X-Airline Policy

A third policy option which is frequently suggested involves replacing the TAP with an X-Airline Policy, where X is usually specified to equal one or three. Monopoly airline advocates[127] stress the economies of scale, economies of aircraft size and scheduling advantages which a single operator could achieve. However, these advantages could be more than offset by the even further reduction in competition and incentive to innovate that this system would entail. In contrast a Three-Airline Policy aims to increase competition in the industry, thus reaping efficiency benefits [e.g. see AFTA (1977, p. 7)]. However, with no threat of further entry the outcome, apart from a possible decline in the incidence of parallel scheduling, may be little different to the present situation. The three operators will face similar incentives to those of the original two, and so there may be little increase in competitive activities. An X-Airline Policy, by interpreting competition in terms of market concentration, offers no guarantee of determining the extent of competitive forces in the industry and hence exerts only a weak influence on economic efficiency.

Deregulation

Several of the policy responses discussed above seem to be based on the premise that an increase in competition in the airline industry would have a favourable impact on economic efficiency. Yet these modifications of existing policy fall well short of achieving the potential efficiency gains. It is increasingly clear that freedom of entry into the industry is the crucial issue of policy reform. Levine (1975, p. 648) notes that ease of entry is more important as a factor in fare competition than even pricing freedom and observes that most examples of price competition and marketing innovations have occurred when entirely new airline firms enter the market. The discussion in previous chapters indicates that open market competition is feasible in the airline

---

127 As noted in Chapter 2 this proposal has a long history of support, starting with the Corbett Committee and most recently advocated by Richardson [see *The Canberra Times*, 11 October 1979] and Brogden (1980).

industry; empirical observation supports this assessment. Relaxation of the only significant entry barrier, the Government's regulatory policy, offers the greatest possibility of a significant improvement in the industry's economic performance. The open market solution provides both the necessary incentive to stimulate management and the most decisive test of their entrepreneurial abilities. Consumers need fear little in a deregulated airline environment; it is the inefficient airline operator who stands most to lose.

The adjustment process

One aspect of the deregulation option which needs further comment is the so-called 'adjustment process'. Some authors are concerned with the possible costs of adjustment which might follow an abrupt deregulation of the industry. BTE (1978, p. 53) thus urges:

> Whatever the regulatory changes need to be, they should be instituted on a gradual basis. Limited experimentation with reformed legislation should be given serious consideration before final decisions are made . . .

Similarly, Forsyth and Hocking (1978, p. 26) recommend:

> . . . judicious shifts away from the present system, not . . . a sudden complete change.

Adjustment costs may be borne by both consumers (e.g. disrupted services) and producers (e.g. adapting the firm to changing market conditions). It is hoped that a phasing-in of policy reform would enable existing operators to increase efficiency and reorient their managerial goals so that they can cope more adequately with the new competitive environment. At the same time the potential for disruptions to services may be lessened.

However, these adjustment costs should not be exaggerated. 'Improved', rather than 'disrupted', may be a more accurate description of the changes in air services which are offered to the majority of consumers. Similarly, the alleged costs to the firm due to the traumatic competitive experience may merely be wealth losses suffered from declining consumer patronage. Demory (1975) argues that the adjustment costs are likely to be small because of the competitive nature of the industry. It was noted in Chapter 1 that the nature of the fixed capital assets of airline operators implied that there were few substantial barriers to entry and exit. Hence the adjustment costs of changing a firm's scale of operations were relatively small. In these circumstances the adjustment costs rationale for a gradual introduction of policy reform may largely reduce to considerations of equity rather than efficiency, i.e. that the beneficiaries of existing policies should not bear most of the burden of changing to a new regulatory framework which is designed to increase economic efficiency.

There may be offsetting dangers in lengthening the period of policy change. A sudden break may be necessary to ensure that the policy reform actually occurs; gradualism may merely represent an attempt to stall the initiatives or to entrench the positions of existing operators. A phasing-in of new policy also presents the authorities with a further regulatory task. Since the policy reform is desired to overcome the inefficiencies produced under the current system, one must be sceptical of the ability of these same regulators to maximise economic efficiency through their discretionary introduction of the new policies. Observation of U.S. deregulation indicates that the inflexibilities, which are inevitably present with a tight system of removal of economic controls (regulated deregulation!), can also have an adverse effect on the industry. Many U.S. airlines have urged more rapid deregulation of the industry. Recent financial difficulties of some airlines have been blamed on their inability, due to regulatory procedures, to adjust rapidly to higher oil prices.[128] Finally, a process of gradual deregulation appears to offer the greatest prospect of predatory pricing in the airline industry. An existing carrier might be able to offset its losses in a deregulated market segment with the profits earned in those still subject to entry restrictions. This possibility needs to be considered in connection with the 'Bizjets incident' and the deregulation of air freight which was recommended by DOT (1979a, p. 118) and which is part of the Government's announced policy initiatives.

From May 1979 to May 1980 Bizjets operated a Metroliner service between Essendon and Devonport in competition with F27 airline services over the Melbourne/Devonport/Wynyard route. The response of the major carriers was to upgrade their operations by providing first and economy class

---

[128] See *Australian Financial Review*, 20 and 26 February 1980.

services and to discount their fares to approximate the new concessional fares being offered by the commuter operator. From 1976 until the introduction of this commuter operation the fare charged for the F27 services was set equal to the economy class jet fare for the Melbourne/Launceston service. Following their discounting the F27 fare represented about 90 per cent of this jet fare. In approving the fare discounts Nixon remarked:

> As Bizjets are now operating over an airline route, it would not be appropriate to constrain the two major operators from competing with the new Bizjet service.[DOT (1979d)]

A few weeks later, on 14 June 1979, the Minister approved increases averaging 10 per cent in the air fares of the major operators. There was no increase in the airlines' fares on the Melbourne/Devonport/Wynyard route, so that this fare now represented only about 80 per cent of the jet fare with which it was previously common-rated. The fare relativity has been restored to some extent since Bizjets ceased operations; in December 1980 this fare was again approximately 90 per cent of the economy jet fare of the Melbourne/Launceston service.

A clear suspicion exists that the demise of the Bizjets operation was significantly affected by predatory pricing perpetrated by the major airlines and condoned by DOT.[129] Two comments can be made. Firstly, this episode is likely to have offended many people's elementary notions of fairness and equity. Secondly, while such predatory pricing in competitive market segments offers little scope for additional profits in a static sense (although it can hardly encourage further market entry), the existence of a history of such company failures would undoubtedly prove useful in the longer run as supposed evidence to support the continuation of existing economic regulation of more significant market segments, i.e. trunk route passenger services.

The above discussion indicates that not only can the alleged costs of adjustment be exaggerated but a protracted period of change can also introduce its own severe problems into the industry. Drawing on his experience with U.S.

---

129 See Government of Tasmania (1980, p. 30) and DOT (1980b, p. 42) for further discussion. The DOT account of this incident does not deny that predatory pricing occurred.

deregulation, Kahn (1979, p. 12) concludes:

> . . . I have as a result been converted to the conclusion that the only way to move is fast. The way to minimise the distortions of the transition, I am now thoroughly convinced, is to make the transition as short as possible.

7

The Choice Between Alternative Institutions: Regulation versus Competition

Conflict in industry policy formulation arises from the incompatibility of regulatory and competitive market forces. Kahn (1971, p. 1) notes:

> the decision to regulate is, typically, a decision also to restrict competition, not just to supplement it in one way or another, but to supplant it.

Hence the decision regarding government involvement in an industry's activities largely amounts to a choice between the institutional frameworks of regulation and competition, and the extent to which each of these forces is permitted to influence the industry. Kahn (1971, p. 46) warns that society should be aware of and take into account the inherent tendencies of these two institutions when choosing among various systems of industrial organisation. This chapter briefly looks at some basic characteristics of regulation and competition[130] and makes a few observations on the choice between the two.

I. SOME TENDENCIES OF THE REGULATORY PROCESS

Conservatism

Many of the recognised virtues of competition, such as the stimulus to innovation and the link between the performance of particular firms and their market fortunes, are often regarded as defects from the regulatory viewpoint. Regulators give emphasis to stability and predictability, qualities which ease their tasks of planning and administration. Consequently regulation tends towards conservatism.

A bureaucratic concern for 'orderly administration' is

---

130 Kahn (1971) discusses many of these issues in greater depth.

evident in the regulation of Australia's airlines. DOT (1979a, p. 31) concludes that the supposed vagueness of definition of its public interest goals does not impede the orderly administration of airline licensing. Further evidence is seen in this Review Committee's preference for organising the industry into a structure containing distinct categories of operators with a sharp definition of roles and strict control of inter-category competition [DOT (1979a, p. 69)]. The bureaucracy has always been able to see a clear distinction between the services offered by the two major airlines and those by other operators. For instance, services of commuter operators are 'not airline services'.[131] While this statement is strictly true within the legalistic setting of Australia's Air Navigation Regulations (commuter operators do not hold 'airline licences'), it is unfortunate if such a rigid approach continues to be applied to the economic analysis and regulation of civil aviation. Similarly, the questionable distinction which was perceived by the Review Committee between trunk and regional operators (based on vague notions of 'community of interest' and responsiveness to needs) should continue to provide sufficient bureaucratic flexibility so as not to hinder orderly administration.

This approach to air services and their operators contrasts with the typical competitive framework advocated by many economists who see little need to attempt to segregate the activities of the various firms willing to offer different services in the market. The advantages claimed for a tightly controlled pattern of industry organisation (e.g. economies of scale, economies of market integration and specialisation, and attention to consumer demands) would also be enjoyed under open market conditions. In addition, one would expect a competitive system to show greater flexibility in adapting to changing market circumstances as well as to provide increased incentives towards efficiency for existing operators from the threat of inter-category competition.

Protectionism

Increased government involvement in an industry through the regulatory processes also increases the government's responsibility for the economic well-being of the industry. Thus regulators may easily see their objectives, and hence self-interest, in terms of the viability of existing firms in the industry. Regulation can then tend towards protectionism,

---

131 DCA, *Annual Report*, 1966/67.

with the authorities becoming increasingly concerned with the problems of the industry, e.g. its marginal profitability, its heavy capital requirements, etc.[132] To some extent then, the regulators can usurp the role of management. Sainsbury notes:

> The way in which you [Mr Weeden] have answered that [question regarding proposed air services] indicates that you see the main problem as the cost to the carriers. You discussed it from the businessmen's point of view, not from the point of view of whether or not you would approve that type of arrangement. It should not interest you terribly much initially whether people are making money or not.[133]

DOT (1979a, p. 69) reveals its paternalistic[134] attitude when it rejects options leading to increased competition among the various categories of operators:

> This carries the danger of operators attempting roles for which they are not ideally equipped.

The possibility of mistaken investments in a competitive market is discussed in Chapter 1. However, the qualifications needed and supposedly possessed by the bureaucracy in order for it to efficiently allocate air transport resources are neither clearly specified nor apparent. Mistakes are also likely on the part of the bureaucracy whose incentive structure may bias its decision-making away from more risky or innovative initiatives. Thus a regulatory environment which denies operators the opportunity to undertake certain activities carries the danger of preventing them from performing roles for which they are ideally equipped.

---

132 E.g. DCA, *Annual Report*, 1968/69.

133 *Hansard, SCT*, 10 May 1978, p. 4716. Weeden is First Assistant Secretary, Air Transport Policy Division, DOT. Many others share a similar dislike for such close concern for the profitability of particular firms; e.g. *The Age*, Editorial, 19 May 1979 considers 'this is none of the Government's business'.

134 Perhaps even more: Anderson (1973, p. 16) apparently considered himself to be a 'benevolent old grandfather' of Australian civil aviation.

Distortion of managerial effort

Management may willingly subject itself to increased bureaucratic guidance and less responsibility in return for a more secure operating environment. For instance, EWA urges:

> Government should take a more active role in defining the job to be done; in establishing the optimal solution; in limiting the number of air services to those which can be justified economically; in providing operators with reasonable security of tenure over their routes.[135]

While the basic self-interest motivation of the firm can be expected to remain unchanged, the regulatory framework, by altering the incentives facing managers, may result in a re-orientation of management goals. With reduced importance given to market mechanisms for assessing a firm's economic performance and deciding its financial fate, management may be able to direct less attention towards cost minimisation and the provision of innovative services and to concentrate on compliance with the regulations and risk minimisation. Menzies was aware of the fundamental requirements for the most efficient development of Australian aviation:

> Of all the means of transport, flying is the one which requires in the highest degree enterprise, a willingness to adventure capital, flexibility of mind, and constant contact with scientific development and commercial practice.[136]

However, Brogden (1968) recognises that throughout its history the real factors determining corporate success in Australian civil aviation have been a 'keen political sense' (p. 11) and 'strong friends in government' (p. 14).

---

135 *Hansard, SCT*, 16 May 1978, p. 4884. This statement is hardly the voice of free enterprise nor does it augur well for those relying on existing regional operators to provide a competitive check and stimulus to the major airlines.

136 *Hansard, HR*, 25 July 1945, p. 4555.

II. IN THE PUBLIC INTEREST?

Airline policy and the public interest

Regulators and others clearly perceive the 'public interest' to be a desired objective; it is often the justification for and stated aim of policy initiatives. Evatt claims:

> . . . government intervention is sought by the Labour movement only in so far as it is intended to prevent exploitation or oppression of the people, or social injustices.[137]

Similarly, Chippindall (1965) has

> . . . no doubt of the single-minded and high motives which have inspired the administrators of that policy over the years.

This interpretation of air transport regulation is supported by Anderson (1973, p. 18):

> Whenever I faced any problem in Civil Aviation, any development, I was always more concerned with the impact of any decision I made on the travelling public, and the good of the public in general. That was my primary motivation . . . We always pressed for things that were, to use a hackneyed phrase, in the public interest.

The Introduction outlined the conditions which are required to ensure that government intervention in an industry improves the economic welfare of the community. Yet this study has revealed little evidence of market failure and has also suggested that the present regulatory system involves a substantial degree of economic inefficiency. Thus, while public interest rhetoric is pervasive in discussions of Australian aviation, sufficient doubt exists regarding the public interest merit of the current policies that an alternative explanation of the choice of these policies should be sought.

One such explanation would emphasise the inability of the authorities to achieve their public interest goals. Regulators may mistakenly interpret the public interest and

---

137 *Hansard, HR*, 29 October 1952, p. 3881.

hence pursue policies which are inimical to that objective. On the other hand they may not have the powers or knowledge to successfully pursue public interest aims. Whatever is the reason, the public interest may not be served. Thus Anderson follows his previous quotation with the comments:

> Also we were motivated by a desire to maintain a viable, efficient air transport industry, because we thought that that was in the public interest, too . . . you had to make a judgement here between what was good for the airlines at any particular time, knowing that you wanted to support them and keep them financially healthy, and what other factors were in the public interest. We fought for years . . .We were only moderately successful. But we tried . . .

The private interest theory of airline policy

Another possible explanation of observed industry policies is provided by the 'private interest' theory of regulation.[138] This theory postulates that regulation is the result of various groups in society (e.g. bureaucrats, airlines and consumers) competing among themselves through the political system to maximise their own welfare. As the effective political lobbying strength varies among such groups, it is possible for some special interest groups to increase their well-being at the expense of the public interest.

What reasons are there to believe that the private interest theory of regulation offers a useful explanation of domestic airline policy? Firstly, it is generally accepted that the economic regulation of Australian aviation has been to the substantial benefit of the two major operators. Ansett's co-partnership in the Two-Airline system is thought a 'prize offering'[139] and 'the major assets dealt in are Government licences and prohibitions upon competition'.[140] Similarly, the bureaucracy enjoys substantial benefits in terms of employment, power and prestige under the existing regulatory arrangements, while Kirby (1980b) notes that a large proportion of the recommendations of DOT (1979a) coincide with increased work opportunities for the admini-

---

138 See Posner (1974) for an overview of the public interest and private interest theories of regulation.

139 *Australian Financial Review* (1961).

140 *Australian Financial Review*, Editorial, 14 December 1979.

strators of Australia's air transport industry.

There is also little doubt that considerable weight is given to the airlines' viewpoint at the policy formulation stage. For instance, Freeland, when questioned on the likelihood of certain innovative pricing arrangements being adopted, answers:

> I do not know what the airlines' reaction would be to that. We were looking particularly at the public interest and it seemed to us that the public interest would be fostered by that. I think I would need to hear the airlines' reaction before I could answer.[141]

In addition, many of the changes in policy which have occurred in recent years may be attributable to shifts in the relative lobbying strengths of particular groups. The Secretary of DOT, Halton (1979) comments:

> . . . certain developments, such as the recent transport policy reviews . . . arose from changes in consumer and community attitudes (p. 1)

and

> Consumer and community attitudes also appear to have changed in recent years possibly due to, on the one hand, a better information flow from governments to consumers and industry and, on the other hand, to various interest groups being better organised and vocal in putting forward their point of view. (p. 4)

While the observation of increased activity from the non-airline groups appears correct, the driving forces are more likely to be better knowledge of the impact of regulation in general, the economic performance of the Australian airlines and the performances of airlines under alternative regulatory frameworks, rather than any government-initiated information flows. The calls for public participation in the regulatory process, which are discussed in the previous chapter, also indicate a belief in the private interest theory of regulation. Public representation gives greater weight than previously given to the interests of groups other than the bureaucracy and the airlines and thus, under the private

---

141 *Hansard, SCT*, 17 August 1978, p. 5593.

interest hypothesis, possibly results in a different regulatory outcome. Finally it was noted in the Nixon quotation in Chapter 4 (p. 56) that, at least with respect to safety matters, aviation policies are often determined on the basis of effective political demand rather than any concept of need or the public interest.

These observations, together with the spurious nature of many of the public interest arguments used to justify airline regulation and the existing degree of economic inefficiency, suggest that the private interest theory of regulation may provide a superior explanation of current airline industry policies and more accurate predictions of future policy changes.

Conclusion

> . . . accuracy always will be a casualty of propaganda but the misapprehensions to which it leads do not make any easier the task of considering logically [airline regulation]. . . or of encouraging informed public debate on it. [Hewitt (1979b, p. 9)]

It has been the aim of this study to expose some of the inaccuracies and misapprehensions which have been present in the debate on the economic regulation of Australian civil aviation. There have been two major areas of discussion. Firstly, the study has examined the issue of market failure in the airline industry. It has investigated the many public interest arguments as to why the market mechanism is an unsuitable method of satisfying the community's demands for air services. These arguments include claims that competition among airlines will be destructive or wasteful, will result in many areas without services, or will lead to monopoly exploitation of consumers. The validity of these arguments is seriously questioned.

Several external benefits alleged to be associated with civil aviation have also been considered. While there is scope for disagreement about the significance of these supposed benefits, it appears that these problems, to the extent they exist, are best dealt with directly through subsidies or other fiscal measures rather than indirectly through restrictive market control. Similarly, it was found that the emotive issue of air safety is largely irrelevant to considerations of the economic regulation of airlines. Air safety objectives are not dependent upon market regulation but are more effectively pursued through direct attention to operating standards and procedures.

Thus it seems clear that the many public interest arguments used in the Australian debate for rejecting market forces and for supporting detailed economic regulation of airlines do not stand close scrutiny.

Secondly, this study examined the economic performance of the industry under the current regulatory framework. In contrast to the the apparent absence of market failure, significant areas of government failure were identified. The Australian air transport industry is marked by a substantial degree of economic inefficiency. A discussion of the policy options available to improve the economic performance of the industry indicated that deregulation, in particular, permitting freedom of entry into the industry, offers the greatest prospects for increased efficiency. In this regard the continued state ownership of TAA not only has little independent impact on the performance of the industry but is, if anything, an obstacle to policy reform.

A theme which is present throughout this study, but especially prominent in the sections analysing the industry's performance and considering various policy options, is the importance of the incentives facing a firm as a determinant of its economic performance. The principal deficiency of the TAP is its failure to provide the major airlines with incentives encouraging them to operate efficiently. Future policy reform should concentrate on correcting this deficiency.

In summary, this study concludes that the combination of a lack of convincing theoretical and empirical arguments to support restrictive airline regulation, the unsatisfactory economic performance of the major airline operators under the current regulatory system and the apparent feasibility of open market competition among airlines indicates the need for substantial regulatory reform.

What is the likelihood of such reform? The private interest theory of regulation suggests that the forces for policy change lie in shifting relative lobbying strengths of the various parties affected by airline regulation. The study noted the increasing influence which consumer-oriented groups have exerted in recent years. In addition, State Governments, especially those of Western Australia and the Northern Territory, have lately taken a higher profile in the industry. It has also been suggested that unless the apparent discrepancy between overseas and domestic air fares is reduced the community will force open competition upon the local operators.[142] However, it is doubtful that these influences are yet strong enough to quickly achieve the

---

[142] Chanticleer, *Australian Financial Review*, 4 June 1979. The differences between domestic and overseas air fares have been clouded recently by frequently changing factor prices, especially oil prices.

degree of reform which this study judges necessary. For example, many dissatisfied consumers, especially those not currently travelling by air, have little opportunity to express their dissatisfaction since there are few close substitutes. Forsyth (1979, p. 63) notes:

> The policy was devised very skilfully such that it would be very difficult to bring any pressure upon it.

In recent years the U.S. domestic airline system has been substantially reformed with policies designed to place increased reliance upon competitive market forces. Many of the elements then stimulating these policy changes are similar to those currently present in Australia, e.g. frequent air fare rises, the embarrassing comparison of the economic performances of the CAB regulated carriers and the less regulated intrastate operators, and advances in economic research of the airline industry. One influential factor not prominent in the Australian situation is the exposure of regulatory malpractices. In the U.S. this issue went further than the general disillusion with government in the post-Watergate period. Kennedy (1975, p. 608)[143] concludes:

> Several of the procedures that the Board follows in setting major Board policies - in particular route and enforcement policies - have lacked the openness, intelligibility, and impartiality required by elementary notions of procedural fairness.

These procedures are also described as 'highly improper and probably illegal'.

It is not certain that the Australian industry has been free of similar bureaucratic abuse. For example, Brogden (1980, p. 119) reports the granting of an air route licence to EWA which, he suggests, was designed to act as a 'big stick' towards the major operators. Government of Western Australia (1980, p. 71) also alleges that DOT has abused its financial and safety powers and the widespread industry acceptance of it as the dominant aviation authority by usurping certain licensing and pricing powers in relation to purely intrastate services. However, the relatively secretive nature of Australian air transport regulation compared with

---

143 This article is the Summary of Report of the Senate Subcommittee on Administrative Practice and Procedure.

the more open, public hearing style of the U.S. system increases the difficulty of detecting undesirable regulatory practices in Australia. The inability of the public to carefully scrutinise Australian regulatory procedures probably lessens the chance of imminent policy reform to the extent that has occurred in the U.S. industry.

Appendix 1

A DESCRIPTION OF AIR TRANSPORT REGULATION IN AUSTRALIA

Several other publications present detailed descriptions of the historical development of Australian aviation policies [e.g. Richardson and Poulton (1968)]. This appendix briefly outlines those powers and policies which determine the present patterns of airline regulation as well as the recently announced proposals for policy amendments.

The impact of the Australian Constitution

Since there is no explicit mention of civil aviation in the Constitution it remains a matter of responsibility for the States. Hence the ability of the Commonwealth to regulate the air transport industry rests with other heads of power, the major ones being its powers to legislate with respect to international and interstate trade and commerce, external affairs, the Territories and Commonwealth places. However, section 92 of the Constitution, which guarantees that trade among the States 'shall be absolutely free', restricts the powers of the Commonwealth to regulate interstate aviation. Thus the authority of the Commonwealth for direct economic regulation of civil aviation is quite limited.

Air Navigation Regulations (ANRs)

The Air Navigation Act 1920-1973 ratifies on behalf of Australia various international agreements on civil aviation and authorises the making of regulations implementing the Act and these agreements and regulations in relation to Territorial, interstate and international air navigation. There are over 300 such ANRs applying to all air navigation within Australia. They mainly deal with technical matters of air operations (e.g. aircraft registration, airworthiness requirements, crew licences, air facilities and flight rules). However, several are important for the economic regulation of domestic air transport.

Licences issued by the Secretary of DOT are required to operate charter and regular public transport air service operations [ANR 197(1) and 198 respectively]. However, there are legal constraints on the assessment of licence applications [ANR 199]. In particular, the only legal grounds for refusing both interstate or intrastate licences relate to safety considerations. There is also an important variation to these licencing requirements. The Secretary may, if he considers it warranted, exempt the holder of a charter licence from the necessity to obtain an airline licence in order to operate scheduled services [ANR 203]. The practical effect of this procedure is to lessen the cost of providing scheduled air services by lowering the technical standards required for these operations. Widespread growth in commuter services has occurred since the introduction of these exemptions in 1967.

Air fares are controlled through ANR 106 which requires the operator of charter or scheduled air services using Commonwealth air facilities to provide the Minister for Transport with information on the fares charged for these services. The Minister can approve the submitted charges, possibly with variations, or can reject them and direct the adoption of fares that he considers 'fair and reasonable'. An airline must also have its timetable approved by the Secretary of DOT [ANR 106C]. However, these timetables can be disapproved only on safety grounds.

The Two-Airline Policy

The so-called Two-Airline Policy (TAP) is the dominant feature of the regulation of Australia's domestic air transport. Since the Government lacks the constitutional authority for direct economic regulation of civil aviation, it has chosen to control entry into the industry and to regulate the behaviour of its selected operators by means of a series of contractual agreements, the various Airlines Agreements Acts. These provide for termination with five years notice by ATI or the Commonwealth. Other features of the policy are embodied in the Airlines Equipment Act 1958, the Australian National Airlines Act 1945 and the Customs (Prohibited Imports) Regulations.

The fundamental policy objective is to maintain only two operators, one being TAA, of trunk route airline services. The Government is able to ensure that trunk route operations are limited to only AAA and TAA through its powers to prohibit the importation of aircraft. It has been Government policy to refuse import permits for all aircraft types over a specified gross weight unless the aircraft is purchased by a present airline operator or unless other restrictive conditions are satisfied.

The total amount of aircraft capacity which is offered by the two major operators is controlled through the Airlines Equipment Act. The capacity determination provisions of this Act require the Minister to estimate future traffic on competitive and non-competitive routes and then, on the basis of a chosen revenue load factor (e.g. 65 per cent for mixed configuration aircraft on trunk routes), to determine the maximum aircraft capacity needed by each airline to cater for its non-competitive route traffic as well as for half of the competitive trunk route traffic. These determinations are made on a six-monthly basis. The airlines are obliged not to provide capacity on trunk routes in excess of their determination. There is no obligation for the airlines to operate the same aircraft type, as witnessed by their recent choices of different wide-bodied jets. However, there is a requirement that new aircraft purchases will not be 'detrimental to the stability of the domestic air transport industry'.

The TAP also allows for the rationalisation of the air services offered by the major operators. The so-called Rationalisation procedures enable TAA and AAA to consult and resolve differences on a wide range of matters, e.g. fares, timetables, aircraft types, capacity and load factors. Thus the airlines have considerable scope for discussion and joint implementation of a wide range of features affecting domestic air services.

The policy also specifies procedures to be followed when the airlines are in dispute. A Rationalisation Committee is established consisting of a representative of each of the airlines and a Co-ordinator nominated by the Minister (the Secretary of DOT). If the airlines are unable to agree on any of the matters which they keep under review, the matter in dispute can be referred to this Committee by either of the airlines. If, after consideration of the issue by the Committee, the airline representatives are still unable to reach agreement the Co-ordinator shall make a final decision. If one of the airlines is dissatisfied with the decision of the Co-ordinator it can appeal to the Arbitrator who is usually a Justice of a federal court. The airline representatives will confer on the matter under the chairmanship of the Arbitrator who, if they are unable to agree, shall decide the issue. The operators are obliged to comply with all their Rationalisation agreements and with any decisions of the Co-ordinator or the Arbitrator.

In response to the problems caused by the preferential treatment of TAA during the immediate post-war years a consistent element of the TAP has been the desire to ensure that competition between the government and private operators is fair and equal. Hence the Commonwealth undertakes not to unfairly discriminate between the two airlines, especially with respect to the granting of import licences and the allocation of airport facilities. Government business is freely available to both operators and each is given a substantially equal share of air mail. The Commonwealth also provides financial assistance in the form of loan guarantees to the private airline to ensure that it can obtain the aircraft necessary to match the services of its competitor. In addition, the Australian National Airlines Act has been amended in an effort to make the cost structures of the two operators as similar as possible. Thus TAA must pay all government taxes and rates, it must keep self-insurance funds in a prescribed account in the form of Commonwealth securities, it must aim to make sufficient profits to meet a dividend target set by the Commonwealth, and it is able to diversify into non-airline activities.

Finally, the airlines face certain obligations under the TAP. These include: not to dispose of aircraft without Commonwealth consent; to maintain existing rural services so long as revenue exceeds direct operating costs and not to cease operating a service without attempting to replace it with an ANR 203 (i.e. commuter) operation; to investigate the introduction of promotional fares and other measures designed to stimulate traffic and air freight; to comply with airport curfews; and to acknowledge that the Commonwealth can permit other operators to develop air services on non-trunk routes and specialist freight and passenger services of a nature not adequately provided for by the major operators.

Proposed amendments of the Two-Airline Policy

In late 1980 the Government announced amendments which it proposed to make to the TAP [see DOT (1980a)]. The Minister for Transport, Mr Hunt, stated that changes to the TAP were necessary 'to foster increased competition within the aviation industry in a rational and orderly manner'. The basic features of the policy amendments include:

(a) removal of air freight from the TAP - Thus freight will be excluded from the capacity determinations of the major operators. Existing and new operators will be able to provide freight services in unrestricted competition.

(b) removal of air mail from the TAP - Australia Post will be able to negotiate without constraint with any domestic carrier in order to fulfil its responsibilities.

(c) precise definition of the roles of air operators - The new Agreement provides extensive definitions of the roles of the trunk airlines and of other operators. The trunk route network is defined. TAA and Ansett will be the only two operators of scheduled domestic passenger services over trunk routes. Other domestic operators can provide services over 'prescribed' routes which are mainly intrastate or current regional routes. There is only limited scope for overlap between the major operators and others. In addition, the respective roles of Qantas and the domestic airlines are defined. The domestic airlines can independently provide only domestic services; Qantas is similarly restricted to international operations.

(d) orderly development of the aviation industry - Import policy will be relaxed to allow the acquisition of large jet aircraft by regional and cargo operators. However, freight operators must assure the Secretary of DOT that new aircraft will be used exclusively for freight operations. Similarly, regional operators wishing to import a large jet aircraft must undertake to comply with the capacity determination procedures of the Airlines Equipment Act. In addition, operators competing with the major airlines over trunk routes will not have approved fares which are lower than those of the major airlines.

(e) revised consultative arrangements - The major airlines 'must consult' only on aircraft utilisation, load factors and core fares. A review system exists for resolving disputes concerning these matters. The airlines 'may consult' on other issues. The 'may consult' provision allows increased scope for unilateral action by an airline.

(f) new procedures for fare approval - The airlines will be required to justify their charges every six months. There will also be a simplified review system for limited cost increases in certain specified areas (e.g. wages, fuel and ANCs).

(g) increased reporting obligations on the major airlines

(h) time horizon of new Agreement -The new Agreement will last for five years after which there will be five years' notice of termination.

Appendix 2

LIST OF ABBREVIATIONS

AAA	Ansett Airlines of Australia
AFCO	Australian Federation of Consumer Organisations
AFTA	Australian Federation of Travel Agents
ALP	Australian Labor Party
ANA	Australian National Airways
ANC	Air Navigation Charge
ANR	Air Navigation Regulation
ATI	Ansett Transport Industries
BPA	Bush Pilot Airways
BTE	Bureau of Transport Economics
CAB	Civil Aeronautics Board
DCA	Department of Civil Aviation
DLP	Democratic Labour Party
DOT	Department of Transport
EWA	East-West Airlines
ITA	Institut du Transport Aerien
TAA	Trans-Australia Airlines
TAP	Two-Airline Policy

References

Aircraft (1961a), 'Mr. Ansett reviews the Two Airline Policy', 40 (11), August, 46-48.

Aircraft (1961b), 'Pattern for controlled competition', 41 (2), November, 24.

Aircraft (1967), 'Change scheduled on Two Airline scene', 46 (5), February, 16-19.

Anderson, D.G. (1958), 'More on Australia's civil aviation policy', *Aircraft*, 37 (4), January, 26-28.

Anderson, D.G. (1969), 'The problems that face DCA', *Aircraft*, 48 (10), July, 24.

Anderson, D.G. (1973), 'It's a new era; a new climate for aviation', *Aircraft*, 53 (2), November, 16-20.

Ansett, R.M. (1965), 'The airline system and its future', *Industrial Progress and Development*, 26 (11), November, 11-15.

Ansett, R.M. (1972), 'Aviation in Australia', *North West Industrial Review*, 4 (5), September/October, 39.

Ansett, R.M. (1976), 'The outlook for the airline industry - The domestic airline view', *Australian Transport*, 18 (9), September, 53-57.

Ansett, R.M. (1979), 'Address to shareholders, 1979', *The Canberra Times*, 22 November, 9.

Ansett Airlines of Australia (1979a), 'The plain facts about domestic air fares', 3 May.

Ansett Airlines of Australia (1979b), 'Australia's Two Airline System', 19 October.

Ansett Transport Industries (1980), 'The regulation of air transport', mimeo.

Ansett Transport Industries Limited, *Annual Report*, various issues.

Australian Federation of Consumer Organisations (1978), 'Competition the key word for airlines', *AFCO Quarterly*, June, 10-12.

Australian Federation of Travel Agents (1977), Presentation to the Domestic Air Transport Policy Review Committee, August.

Australian Financial Review (1961), 'Ansett-ANA - The inside story', 20 July, 14-19.

Australian Transport (1977), 'Transport and defence', 19 (11), November, 17-19.

Barnett, A., Abraham, M. and Schimmel, V. (1979), 'Airline safety: some empirical findings', *Management Science*, 25 (11), November, 1045-1056.

Brenner, M.A. (1975), 'Need for continued economic regulation of air transport', *Journal of Air Law and Commerce*, 41, 793-813.

Brogden, S. (1968), *Australia's Two-Airline Policy*, Melbourne University Press.

Brogden, S. (1980), 'Fly United', *Australian Penthouse*, May, 116-121.

Bureau of Transport Economics (1978), *Australian Domestic Air Transport - Some Economic Issues*, Occasional Paper 15, AGPS, Canberra.

Bush Pilot Airways (1977), Submission to the Domestic Air Transport Policy Review.

Butler, C.A. (1971), *Flying Start*, Edwards and Shaw, Sydney.

Callison, J.W. (1975), 'Airline deregulation - a hoax?', *Journal of Air Law and Commerce*, 41, 747-791.

Chippindall, G. (1965), 'Why I believe in the Two-Airline plan', *Australian Financial Review*, 14 September, 2-3.

Commonwealth of Australia, *House of Representatives, Select Committee on Tourism*, AGPS, Canberra, various issues.

Commonwealth of Australia, *Parliamentary Debates, Senate* and *House of Representatives*, AGPS, Canberra, various issues.

Confederation of Australian Industry (1980), Submission to the Independent Public Inquiry into Domestic Air Fares, 4 September.

Corbett, D. (1965), *Politics and the Airlines*, Allen and Unwin, London.

Davies, D.G. (1971), 'The efficiency of public versus private firms, the case of Australia's Two Airlines', *Journal of Law and Economics*, 14 (1), April, 149-165.

Davies, D.G. (1977), 'Property rights and economic efficiency - the Australian airlines revisited', *Journal of Law and Economics*, 20 (1), April, 223-226.

Davies, D.G. (1980), 'Property rights and efficiency in a regulated environment', *Economic Record*, 56 (153), June, 186-189.

Demory, W.L. (1975), 'Deregulation, the adjustment process', *Journal of Air Law and Commerce*, 41, 873-883.

Demsetz, H. (1968), 'Why regulate utilities?', *Journal of Law and Economics*, 11, 55-65.

Department of Transport, *Australian Transport (Annual Report)*, AGPS, Canberra, various issues.

Department of Transport (1979a), *Domestic Air Transport Policy Review*, Volume 1: Report, AGPS, Canberra.

Department of Transport (1979b), *Domestic Air Transport Policy Review*, Volume 2: Appendixes, AGPS, Canberra.

Department of Transport (1979c), *News Release*, 'Proposed BPA operation over TAA's F27 routes in Queensland', 2 February.

Department of Transport (1979d), *News Release*, 'Reduced air fares to North West Tasmania', 24 May.

Department of Transport (1979e), 'Air Transport Industry -

Cost Recovery', Appendix A11.1 of Department of Transport (1979b).

Department of Transport (1980a), *News Release*, 'Australian National Airlines Act. Second Reading Speech by the Minister for Transport', 18 September.

Department of Transport (1980b), Submission to the Independent Public Inquiry into Domestic Air Fares, December.

Dignam, A.B. (1979), 'An answer to the critics', *Australian Transport*, 21 (7), July, 9-10.

Edwards, R., Chairman (1969), *British Air Transport in the Seventies*, Report of the Committee of Inquiry into Civil Air Transport, May, HMSO, London.

Forsyth, P.J. (1979), 'The Two Airline Policy: its results and future', *The Australian Quarterly*, 51 (3), September, 62-73.

Forsyth, P.J. and Hocking, R.D. (1978), 'Economic efficiency and the regulation of Australian air transport'. Paper presented at Seventh Conference of Economists, Sydney, August 28 - September 1.

Forsyth, P.J. and Hocking, R.D. (1980), 'Property rights and efficiency in a regulated environment: The case of Australian airlines', *Economic Record*, 56 (153), June, 182-185.

Freeland, C.W. (1977), 'The regulatory and policy functions of the Air Transport Policy Division of the Department of Transport and the current Review of Domestic Air Transport Policy'. Address to Ballarat College of Advanced Education, Ballarat, N.S.W., 29 September.

Gannon, C.A. (1979a), 'Parallel Scheduling', Appendix A12.1 of Department of Transport (1979b).

Gannon, C.A. (1979b), 'Pricing of Domestic Airline Services - Selected Aspects of Fares on Australia's Competitive Routes', Appendix A10.1 of Department of Transport (1979b).

Gillison, D. (1965), 'The "Two-Airline" Policy : A special report. Parts 1-5', *Aircraft*, 44 (6-9, 11).

Goodrich, R.K. (1960), 'The economic structure of interstate air transport in Australia (1921-1958), Master of Commerce Thesis, University of Melbourne.

Government of Tasmania (1980), Submission to the Independent Public Inquiry into Domestic Air Fares, September.

Government of Western Australia (1980), Submission to the Independent Public Inquiry into Domestic Air Fares, DGT Report No. 254, September.

Grenning, M. and Coat, J. (1979), 'The natural monopoly argument for regulation: Does it apply to Australia's domestic airlines'. Paper presented at Eighth Conference of Economists, La Trobe University, 27-31 August.

Halton, C.C. (1979), 'Current issues confronting Australian transport - The Department of Transport viewpoint'. Address to Chartered Institute of Transport, Canberra, 11 September.

Hartley,W.D. (1974), 'The friendly skies : Tweedledum and Tweedledee airlines divide Australian market and money', *Wall Street Journal*, 19 June, 36-37.

Hewitt, L. (1979a), 'Some aspects of international deregulation'. Address to the Wings Club, New York, 11 April.

Hewitt, L. (1979b), 'A world airline perspective and deregulation - Quo vadis?'. Address to an International Aviation Conference organised by Lloyd's of London Press, New York, 25 May.

Hillyar, K.N.M. (1969), 'Airline operation - today and tomorrow', *Australian Transport*, 13 (4), March/April, 17-21.

Hocking, D.M. and Haddon-Cave, C.P. (1951), *Air Transport in Australia*, Angus and Robertson, Sydney.

Hocking, R.D. (1972), *Some Economic Aspects of Australia's Two Airline Policy*, CEDA M Series No. 35, Council for the Economic Development of Australia, Melbourne.

Hocking, R.D. (1979a), 'The Economic Efficiency of the

Australian Airlines', Appendix A6.2 of Department of Transport (1979b).

Hocking, R.D. (1979b), 'Alternatives to the Two Airline Policy', Appendix A9.1 of Department of Transport (1979b).

Hotelling, H. (1929), 'Stability in competition', *Economic Journal*, 39, March, 41-57.

Institut du Transport Aerien (1979), *Impact of Deregulation on the Air Transport System*, Progress Report No.1: 'The present impact of the regulatory reform on the U.S. trunk airline productivity and profitability', September, Paris.

Jones, C.K. (1975), 'Air transport - Australia's lucky transport mode', *Aircraft*, 55 (2), November, 18.

Khan, A.E. (1971), *The Economics of Regulation: Principles and Institutions*, Volume 2, John Wiley and Sons, New York.

Kahn, A.E. (1979), 'Applications of economics to an imperfect world', *American Economic Review*, 69 (2), May, 1-13.

Kennedy, E.M. (1975), 'Airline regulation by the Civil Aeronautics Board', *Journal of Air Law and Commerce*, 41, 607-636.

Keplinger,B. (1976), 'An examination of traditional arguments on regulation of domestic air transport', *Journal of Air Law and Commerce*, 42, 187-212.

Kirby, M.G. (1979), 'An economic assessment of Australia's Two Airline Policy', *Australian Journal of Management*, 4 (2), October, 105-118.

Kirby, M.G. (1980a), 'Optimal pricing policies for air transport networks: Some comments on economic regulation and the Australian domestic air industry', *Transportation Research*, 14A(4), August, 275-278.

Kirby, M.G. (1980b), 'A critical examination of the Domestic Air Transport Policy Review', A submission to the Independent Public Inquiry into Domestic Air Fares, November.

Levine, M.E. (1965), 'Is regulation necessary? California air transportation and national regulatory policy', *Yale Law Journal*, 74(8), July, 1416-1447.

Levine, M.E. (1975), 'Alternatives to regulation: competition in air transportation and the Aviation Act of 1975', *Journal of Air Law and Commerce*, 41, 703-726.

Lloyd-Jones, D.J. (1975), 'Deregulation and its potential effect on airline operations', *Journal of Air Law and Commerce*, 41, 815-841.

Mackay, K.R. (1979), 'A Comparison of the Relative Efficiency of Australian Domestic Airlines and Foreign Airlines', Appendix A6.1 of Department of Transport (1979b).

McKenzie, R.B. and Tullock, G. (1978), *Modern Political Economy*, McGraw-Hill, New York.

Miller, J.C. (1975), 'A perspective on airline regulatory reform', *Journal of Air Law and Commerce*, 41, 679-701.

Morris, P. (1980a), 'Domestic aviation in Australia', Address, Mount Isa, 28 March.

Morris, P. (1980b), *Press Statement*, 'Liberal Council supports sale of TAA', 13 April.

Nixon, P.J. (1976), 'Airport management in Australia', *Australian Transport*, 18 (9), September, 39-43.

Nixon, P.J. (1978), 'Airlines', *Australian Transport*, 20(9), September, 13.

Perry, J. (1975), 'Civil Aviation at the crossroads', *AOPA Magazine*, 28 (12), December, 54-60.

Posner, R.A. (1974), 'Theories of economic regulation', *Bell Journal of Economics*, 5 (2), 335-358.

Poulton, H.W. (1959), 'Legal and policy aspects of air transport in Australia', *Journal of Air Law and Commerce*, 26 (1), 13-35.

Poulton, H.W. (1965), 'Does regulation stifle progress in transport? - Air transport', *Australian Transport*, 9, October/December, 15-16.

Qantas Airways (1980), Submission to the Independent Public Inquiry into Domestic Air Fares, November.

Ramsden, J.M. (1976), *The Safe Airline*, Macdonald and Jane's, London.

Rasenberger, R.J. (1975), 'Deregulation and local airline service - An assessment of risks', *Journal of Air Law and Commerce*, 41, 843-872.

Richardson, J.E. (1965), 'Aviation law in Australia', *Federal Law Review*, 1, June, 242-291.

Richardson, J.E. and Poulton, H.W. (1968), 'Australia's Two-Airline Policy - Law and the layman', *Federal Law Review*, 3, June, 64-85.

Ryland, J.P. (1967), 'Looking back - and forward', *Aircraft*, 46(9), June, 24-25.

Smith, A. (1978), 'Domestic airline regulation: The case of Australia's Two Airline Policy', Honours in Economics Thesis, University of Queensland.

Swartz, R.W. (1967), 'Trans-Australia Airlines - the Government's view', *Aircraft*, 46 (9), September, 23.

Swartz, R.W. (1969), 'The role of DCA', *Aircraft*, 48 (10), July, 23.

Thayer, F.C. (1977), 'And now "the deregulators" : When will they learn?', *Journal of Air Law and Commerce*, 43, 661-689.

Trans-Australia Airlines (1978), 'Air fares - a comparison. TAA Survey', *This Air Age*, July.

Trans-Australia Airlines, *Annual Report*, various issues.

Trans-Australia Airlines (1980), 'Two-Airline competition is fair', *Information from Trans-Australia Airlines*, 20 January.

Walsh, H.F. (1951), 'A case for total private enterprise airlines', *Aircraft*, 29(12), September, 44.

Warner, J.M. (1971), 'Domestic airlines survey', *Aircraft*, 51(1), October, 46.

Watkins, J.L. (1963), 'Australia's internal air transport', *Journal of the Royal Aeronautical Society*, 67 (625), January, 11-36.

Wettenhall, R.L. (1962), 'Australia's Two-Airline system under review', *The Australian Quarterly*, 34 (1), March, 36-46.

Wettenhall, R.L. (1966), 'Politics and the airlines', *Public Administration*, 25 (1), March, 68-74.

Wilson, J.L.J., Ed. (1962), 'Australian airlines and a rational policy', *Current Affairs Bulletin*, 30 (3), June 18, 34-48.